AF886392

The Art of Product Management

Building the Bedrock for Product Success

Steve Rugh

Jester Labs LLC

Paperback ISBN: 978-1-961658-00-4

Hardcover ISBN: 978-1-961658-01-1

eBook ISBN: 978-1-961658-02-8

First edition 2023

Contents

Preface

There is an art to every discipline in business and life, and the field of product management is no different. It is a world that harmoniously blends scientific precision and the creative allure of art. An arena where the dreamers and the doers, the visionaries and the strategists, the artists and the architects convene to transform the abstract into the tangible outcomes that can transform the world. Welcome to "The Art of Product Management: Building the Bedrock for Product Success."

I bring 15 years of experience in product management, including over a decade spent leading product teams. The better part of my career has been with one of the most revered tech giants – Microsoft. My journey with this prestigious organization offered me the incredible opportunity to work on two of the most prominent applications in the world - Microsoft Office and the Microsoft Edge browser based on Chromium. These experiences have endowed me with invaluable insights into the complex universe of product management.

Throughout my journey, I often witnessed individuals in leadership positions venturing into opinionated directions, armed solely with enthusiasm and lacking a foundational basis in product strategy or vision. This disturbing pattern inspired me to write "The Art of Product Management." The book aims to fill this alarming void and equip product managers with the fundamental principles that underpin this intricate discipline.

With its unopinionated stance, this book delves into three cornerstone aspects—Product Vision, Product Strategy, and Product Roadmaps. In my experience, these form the bedrock upon which great products are built and successful businesses are established. By mastering these facets, you can begin to navigate the vast ocean of product management with confidence and poise.

Product Vision is the guiding star, the dream you aspire to transform into reality. It is what instills your product with purpose and direction. Product Strategy, on the other hand, is your compass. It helps you chart the best course to reach your destination, making key decisions on what to prioritize and when. Lastly, the Product Roadmap is your journey's logbook. It's a living, breathing document providing a step-by-step guide to bringing your vision to life.

The anecdotes and stories woven into this book emerge from my own experiences, offering you insights and practical wisdom from years in the trenches. You'll find tales of victories, hard lessons learned from failures, and the invaluable insights they collectively offered.

However, this book is just the beginning. In my blog and future newsletters and books, I will leverage the understanding you gain here to offer more opinionated perspectives and advanced insights into the world of product management.

Whether you are an aspiring product manager eager to learn the ropes, a seasoned veteran seeking a fresh outlook, or someone just curious about the field, I invite you to delve into the pages of "The Art of Product Management." My hope is that it will help you craft compelling visions, devise impactful strategies, build effective roadmaps, and create products that leave an indelible mark on the world.

Let's embark on this journey to demystify the art of product management.

-Steve Rugh (@steverugh)

Ground Zero: Beginning Your Product Journey

"Vision without action is merely a dream. Action without vision just passes the time. Vision with action can change the world." - Joel A. Barker

Welcome to be beginning of your journey through the exciting world of product management! We'll start our journey by laying the foundations for creating exceptional products that customers love. This section will help you understand the critical aspects of crafting a compelling product vision, defining your product strategy, and building and managing a product roadmap.

In this introduction, we'll summarize the following key topics:

— The power of a well-defined product vision and how it inspires your team, aligns your organization, and sets the stage for success.

— How to create a clear and actionable product strategy that guides your team's decision-making process and ensures your product is on the right track.

— The importance of a robust product roadmap in communicating priorities, aligning stakeholders, and managing expectations.

Let's explore why product vision, strategy, and roadmap are essential to your success as a product manager.

Product Vision (The "Why")

Product vision is the foundation for everything you do as a product manager. It helps you answer the critical question, "Why does our product exist?" By articulating the purpose of your product and the impact it aims to create, the vision statement gives your team a clear direction and a shared understanding of what you're working towards. This clarity helps foster alignment, motivation, and commitment among team members and the organization.

A strong product vision also provides a guiding framework for decision-making. When faced with tough choices or competing priorities, your product vision serves as a compass, helping you evaluate options and make decisions that align with your long-term goals.

A well-defined product vision is essential for several reasons:

Inspiration: A powerful vision statement can inspire your team and spark creativity, motivating them to go above and beyond in their work. For example, Tesla's vision of a sustainable energy future has inspired many employees to work tirelessly to revolutionize the automotive and energy industries.

Alignment: A clear product vision helps align your organization's efforts by providing a shared understanding of the purpose and goals of the product. For instance, Google's early vision to "organize the world's information and make it universally accessible and useful" helped focus the company's resources on creating innovative search and information management solutions.

Decision-making: A strong product vision enables you to make better decisions supporting your long-term goals. For example, Apple's commitment to creating "insanely great" products has guided the company's design and development choices, resulting in many iconic and successful products.

Product Strategy (The "What")

Product strategy helps you answer the question, "What do we need in order to achieve our product vision?" It provides a roadmap for turning your vision into reality – outlining the specific goals, objectives, and initiatives to drive your product's success. By defining your target market, understanding customer needs, and analyzing the competitive landscape, your product strategy ensures that your product is uniquely positioned to create value and meet user expectations.

A well-crafted product strategy also provides a framework for prioritization and resource allocation. It helps you focus on the most critical initiatives, ensuring that your team's efforts align with your product's long-term success and that you use your limited resources best.

A well-defined product strategy is vital for several reasons:

Prioritization: A solid product strategy helps you prioritize features and initiatives based on their potential impact, ensuring that your team works on the most worthwhile projects. For instance, when Amazon started, they focused on selling books before gradually expanding to other categories, prioritizing their efforts based on market opportunities and customer needs.

Resource allocation: With a clear product strategy, you can make better decisions about allocating resources, such as budget, personnel, and time, to maximize your product's success. For example, when Netflix invested in producing original content, it strategically allocated resources to create high-quality shows and movies that would attract and retain subscribers.

Competitive advantage: An effective product strategy helps you identify your unique value proposition and distinguish your product from competitors. For example, Spotify's product strategy has focused on personalized music discovery and seamless user experience, which has helped the platform stand out in the crowded music streaming market.

Product Roadmap (The "When")

The product roadmap helps you answer the question, "When will we deliver on our product strategy?" It visualizes your product's strategic plan, outlining the key features, initiatives, and milestones you'll need to achieve over time. The roadmap helps you communicate priorities, set expectations, and track progress within your team and with external stakeholders.

An effective product roadmap also provides a powerful tool for managing uncertainty and change. By maintaining a flexible and adaptable roadmap, you can respond to shifting market dynamics, customer feedback, and internal priorities while staying focused on your product vision and strategy.

A robust product roadmap is crucial for several reasons:

Communication: A product roadmap helps you communicate your plans and priorities to stakeholders, including team members, executives, investors, and customers. You can effectively manage expectations and maintain transparency by clearly outlining your product's direction and timeline.

Coordination: A roadmap is a tool helping different teams within your organization align their efforts around shared goals and timelines. For example, a product roadmap can help coordinate the work of development, design, marketing, and sales teams to ensure a successful product launch.

Adaptability: A flexible product roadmap allows you to adapt to changing market conditions, customer feedback, and new opportunities. For instance, in response to the COVID-19 pandemic, many companies adjusted their product roadmaps to prioritize features and initiatives that supported remote work and online collaboration.

Now that we've scratched the surface of the importance of product vision, strategy, and roadmap, let's dive deeper into each of these concepts, where we'll explore each of these critical elements in detail. Let's get started!

Chapter 1

The Guiding Star: Unleashing Your Product Vision

"The only thing worse than being blind is having sight but no vision."
- Helen Keller

As you embark on this journey, let us first unveil the curtain and reveal the essential ingredient that gives life to every successful product: the product vision. Like Dorothy in The Wizard of Oz, you may discover that the answers you seek are within you all along. But, to help you on this quest, we will break down the importance of product vision and the critical steps to crafting a captivating one.

A product vision is the North Star that guides your product and your team. It provides a clear direction, illuminates the path forward, and keeps everyone on track. It's not about what your product is today but about what it could become in the future. It's the dream you're chasing, the goal you're striving for, and the purpose that drives every decision you make.

The following chapters will explore the importance of product vision and discuss the elements that make a great vision.

We will provide practical tips and techniques to craft your product vision statement that resonates with your team and users.

- The Essence of a Product Vision
- Crafting Your Product Vision Statement
- Aligning Your Team and Organization Around the Vision
- Evolving Your Product Vision Over Time
- Communicating Your Vision to Stakeholders

Now, let's dive into the world of product vision and unlock its secrets to building great products.

Chapter 2

The Essence of a Product Vision

"Your vision will become clear only when you can look into your own heart. Who looks outside, dreams; who looks inside, awakes." - Carl Jung

At its core, a product vision is the guiding light that illuminates the path for your product and team. It's the dream you're chasing, the goal you're striving for, and the purpose that drives every decision you make. This chapter will explore the essence of a product vision and its role in product management.

A product vision serves several vital functions, such as:

Inspiration: A powerful product vision inspires your team and gives them a reason to believe in your product. It creates a shared purpose and a sense of excitement about the future. By providing a compelling "why," your product vision keeps your team motivated and engaged, even during the most challenging times.

Direction: A clear product vision helps your team make better decisions by providing a long-term perspective. It acts as a compass that points the way forward and ensures everyone moves in the same direction. When faced with multiple options or trade-offs, your product vision can be a useful reference point to guide your team's choices.

Focus: Your product vision helps your team focus on what's truly important. It acts as a filter to prioritize features, initiatives, and investments that align with your vision and contribute to your product's long-term success. By keeping your team focused on the big picture, your product vision prevents distractions and effectively allocates your resources.

Alignment: A well-articulated product vision promotes alignment across your organization, from engineering and design to sales and marketing. By creating a shared understanding of what you're trying to achieve, your product vision helps different departments work together more effectively and reduces friction caused by competing priorities or misaligned goals.

Differentiation: Your product vision sets your product apart from the competition by articulating what makes it unique and valuable. It helps you identify and capitalize on your product's unique strengths, carving out a distinctive position in the market.

A successful product vision is rooted in deeply understanding your users, their needs, and the market. It reflects your core values and beliefs as a foundation for your product strategy and roadmap. By keeping your product vision front and center, you'll be better equipped to navigate the complex world of product management and steer your product toward success.

Let's dive deeper into each of these five crucial aspects of a product vision:

Inspiration

A product vision has the power to inspire not only your team but also your customers, stakeholders, and even yourself. An inspiring product vision can:

Ignite passion: A compelling vision sparks enthusiasm and commitment, driving your team to push through challenges and remain persistent in their pursuit of success.

Create emotional connections: An inspiring vision can forge a strong emotional bond between your product and users by resonating with your target audience's core desires and values.

Foster a sense of purpose: A shared vision helps everyone understand their role in achieving a common goal, fostering a sense of meaning and purpose in their work.

Direction

A clear product vision provides essential guidance for your team, enabling them to make better decisions and navigate challenges. Direction from a product vision can:

Establish priorities: By highlighting long-term objectives, a product vision helps your team prioritize tasks and allocate resources more effectively.

Inform decision-making: When confronted with trade-offs or competing options, a product vision serves as a touchstone, guiding your team to make choices that align with your goals.

Maintain consistency: A well-defined vision helps ensure your product's features and messaging remain consistent, creating a cohesive and harmonious user experience.

Focus

A strong product vision helps your team focus on your product's most critical aspects, preventing distractions and maximizing productivity. Focus from a product vision can:

Keep the big picture in mind: By emphasizing the overarching vision, your team is less likely to get bogged down in minutiae and more likely to stay focused on the long-term objectives.

Eliminate unnecessary features: A clear vision helps your team identify and prioritize features that contribute to the product's success while avoiding those that don't align with your vision.

Streamline decision-making: A product vision can simplify decision-making and help your team focus on what truly matters by serving as a reference point.

Alignment

A well-articulated product vision promotes alignment across your organization, fostering collaboration and synergy between different teams. Alignment from a product vision can:

Unify the organization: A shared vision creates a common understanding of the product's goals and purpose, encouraging cross-functional collaboration and minimizing silos.

Enhance communication: By providing a consistent and clear message, a product vision helps ensure everyone is on the same page, reducing misunderstandings and miscommunications.

Optimize resource allocation: A shared vision helps different departments prioritize their efforts and resources, ensuring everyone works together towards a common goal.

Differentiation

Your product vision sets your product apart from the competition, highlighting its unique value proposition and positioning it for success in the marketplace. Differentiation from a product vision can:

Identify competitive advantages: A compelling vision helps you recognize and leverage your product's unique strengths, enabling you to outperform your competitors and capture market share.

Appeal to your target audience: A well-defined vision ensures that your product speaks directly to the needs and desires of your target users, making it more attractive and relevant to them.

Foster brand loyalty: By articulating a distinct and consistent vision, your product can establish a strong brand identity that resonates with users, fostering long-term loyalty and trust.

By understanding and emphasizing these five aspects of a product vision, you can harness its full potential to inspire, guide, and focus your team, align your organization, and differentiate your product in the market.

Key Takeaways

- A product vision provides inspiration, direction, focus, and differentiates your product.
- The vision helps guide your team's decisions and maintains consistency.
- Promotes cross-functional collaboration and efficient resource allocation.
- Identifies unique strengths and appeals to the target audience.
- Ignites passion, creates emotional connections, and fosters a sense of purpose.
- Success relies on understanding your users, their needs, and the market.

Chapter 3

Crafting the Future: Your Product Vision Comes Alive

"To accomplish great things, we must not only act but also dream; not only plan but also believe." - Anatole France

Your product vision statement is a concise, memorable expression of your product's ultimate goal and the impact you want it to have on the world. Creating a powerful product vision statement is crucial in defining your product's direction and purpose. This section will explore the key elements of a great product vision statement and provide guidelines to help you craft your own.

User-centric: Your product vision statement should focus on the needs and desires of your target users. It should be deeply rooted in understanding your audience, their problems, and how your product will address them. You'll create a more compelling and meaningful statement by putting users at the heart of your vision.

Clear and concise: A great product vision statement is easy to understand and communicate. It should be simple enough for anyone in your organization to grasp and explain. Aim to convey the essence of your vision in just a sentence or two, using clear, straightforward language.

Inspiring: Your product vision statement should evoke a sense of excitement and possibility. It should paint a vivid picture of the future you're striving to create, capturing your team's and stakeholders' imagination. Use evocative language and powerful imagery to make your vision statement as inspiring as possible.

Long-term perspective: A product vision statement should look beyond the immediate future and consider the long-term impact of your product. It should provide a guiding star for your team to follow, even as specific goals and strategies evolve.

Unique and memorable: Your product vision statement should be distinctive, setting your product apart from competitors and capturing its unique value proposition. Aim to create a memorable message that resonates with your target audience and leaves a lasting impression.

Here are some steps to help you craft your product vision statement:

Understand your users: Research your target audience, their needs, and their problems. Identify the most pressing issues that your product will address.

Define your product's purpose: Determine the core purpose of your product and the impact you want it to have on your users and the world.

Brainstorm key phrases and concepts: Generate a list of words, phrases, and ideas that capture the essence of your product vision. Consider what sets your product apart and what makes it unique.

Draft and refine: Combine key concepts to create a concise, powerful statement. Experiment with different phrasings and wordings to find the most compelling expression of your vision.

Test and iterate: Share your draft product vision statement with your team, stakeholders, and target users. Gather feedback and refine your message as needed to ensure it resonates with your audience and effectively communicates your vision.

By crafting a clear, user-centric, inspiring, and memorable product vision statement, you'll provide a solid foundation for your product's success and a clear sense of direction for your team.

Let's dive deeper into each of these five critical elements of a great product vision statement:

User-centric

A user-centric product vision statement revolves around the needs and desires of your target users. To create a user-centric vision:

Conduct user research: Speak to your target users, observe their behavior, and gather insights into their needs, desires, and pain points.

Identify key user benefits: Determine the most important benefits your product will offer to users and how it will solve their problems or fulfill their needs.

Reflect user needs in your vision: Ensure your vision statement clearly communicates the value your product will deliver to users and how it will improve their lives.

Clear and concise

A clear and concise product vision statement is easy to understand and communicate. To achieve this:

Use plain language: Avoid jargon, buzzwords, and overly complex language. Use simple words and phrases that are easily understood by a broad audience.

Keep it short: Aim for one or two sentences that capture the essence of your vision. Resist the temptation to include too much detail or explain every aspect of your product.

Focus on clarity: Make sure your vision statement is unambiguous and leaves no room for misinterpretation. It should be crystal clear to anyone who reads it.

Inspiring

An inspiring product vision statement evokes a sense of excitement and possibility. To make your vision statement inspiring:

Use vivid imagery: Paint a vivid picture of the future you're striving to create with your product. Use descriptive language and powerful images that evoke emotion and capture the imagination.

Connect with emotions: Your vision statement should resonate emotionally with your team, users, and stakeholders. Appeal to their values, aspirations, and sense of purpose.

Be bold and ambitious: Don't be afraid to dream big and set high aspirations for your product. An inspiring vision should challenge and push your team to reach their full potential.

Long-term perspective

A product vision statement should take a long-term perspective, looking beyond immediate goals and considering the lasting impact of your product. To incorporate a long-term view:

Focus on the ultimate goal: Think about the result you want to achieve with your product rather than getting bogged down in short-term objectives or specific features.

Be flexible and adaptable: Acknowledge that the path to realizing your vision may change over time. Your vision statement should be broad enough to accommodate strategy, technology, and market conditions shifts.

Consider lasting impact: Contemplate the legacy you want your product to leave behind and how it will contribute to the greater good in the long run.

Unique and memorable

A unique and memorable product vision statement sets your product apart from competitors and captures its distinctive value proposition. To make your vision statement unique and memorable:

Emphasize differentiation: Highlight the aspects of your product that make it unique and distinguish it from competitors. Focus on the specific value proposition that sets your product apart.

Be creative: Use imaginative language, metaphors, or analogies to convey your vision in a memorable way. Be bold and think outside the box and break away from conventional phrasing.

Make it memorable: Aim for a catchy, unique statement that sticks in people's minds. It should be something that your team, users, and stakeholders can readily recall and share with others.

By focusing on these five elements—user-centric, clear and concise, inspiring, long-term perspective, and unique and memorable—, you can create a strong product vision statement that effectively communicates your product's purpose and sets the stage for its success.

Let's look at examples of excellent and poor product visions and explain why each falls into its respective category.

Great Product Vision Examples:

Tesla: To accelerate the world's transition to sustainable energy.

- This vision is user-centric, focusing on the global need for sustainable energy.

- It's clear, concise, and easy to understand.
- The vision is inspiring, bold, and ambitious.
- It takes a long-term perspective, considering the lasting impact of sustainable energy.
- Tesla's unique and memorable vision sets them apart from traditional automakers.

Microsoft: To empower every person and every organization on the planet to achieve more.

- The vision is user-centric, emphasizing empowerment for both individuals and organizations.
- It's clear, concise, and communicates the intent effectively.
- The statement is inspiring and appeals to a sense of purpose and aspiration.
- It takes a long-term perspective, as empowering people and organizations is an ongoing endeavor.
- Microsoft's vision is unique and memorable, focusing on empowerment rather than just technology.

Airbnb: To create a world where anyone can belong anywhere.

- This user-centric vision focuses on travelers' need for

belonging and connection.

- It's clear, concise, and effectively communicates the intent.
- The statement is inspiring, with strong emotional resonance.
- It takes a long-term perspective, envisioning a more inclusive world.
- Airbnb's unique and memorable vision sets them apart from traditional hospitality businesses.

Poor Product Vision Examples:

To build a platform that enables businesses to grow and succeed.

- This vision is generic and could apply to many different products or companies.
- It needs a clear focus on user needs and could be more inspiring.
- The statement doesn't showcase any unique or memorable aspects of the product.

To leverage cutting-edge technologies to revolutionize the way people interact with digital content.

- This jargon-heavy vision makes it harder for a broader audience to understand.
- It focuses more on technology rather than the user's needs or experiences.
- The statement is neither inspiring nor memorable, lacking a clear long-term perspective.

To become the leading provider of innovative solutions in the global market for XYZ.

- This vision must be more specific and communicate a clear purpose or user benefit.
- It could be more inspiring, focusing on the company's goals rather than the impact on users.
- The statement is not unique or memorable and lacks a long-term perspective.

These examples show that excellent product vision statements are user-centric, clear, concise, inspiring, long-term focused, and unique. In contrast, poor ones lack these qualities and often focus on company-centric goals, use jargon, or remain too generic.

Key Takeaways

- A product vision statement is a succinct, memorable expression of a product's ultimate goal.
- Essential elements include being user-centric, clear, inspiring, long-term, and unique.
- Crafting this statement involves understanding users, defining the product's purpose, brainstorming key phrases, and refining the statement based on feedback.
- Great product vision statements are exemplified by those from Tesla, Microsoft, and Airbnb.

Chapter 4

Building Consensus: Aligning Your Team with the Vision

"Individual commitment to a group effort—that is what makes a team work, a company work, a society work, a civilization work." - Vince Lombardi

Creating a compelling product vision is just the beginning. The real challenge lies in aligning your team and organization around that vision. When everyone is on the same page, it fosters a sense of unity, promotes efficient decision-making, and increases the likelihood of success. This chapter will explore strategies for aligning your team and organization around your product vision.

Communicate the Vision Clearly and Frequently

Effective communication is the cornerstone of successful alignment, and it begins with expressing your product vision clearly and engagingly. To ensure that your team and organization understand and embrace the vision, it's essential to communicate it regularly and through various channels. Here are some tips to help you share the vision clearly and frequently:

Be consistent in your messaging: Use consistent language, terminology, and imagery when discussing the product vision. This will help reinforce the message and ensure everyone is on the same page.

Tailor your communication to your audience: Different team members may have varying levels of understanding and interest in the product vision. Adjust your communication style to cater to different audiences, using language and examples that resonate with them.

Leverage various communication channels: Use multiple channels, such as meetings, presentations, emails, and internal blogs, to convey the vision. This helps ensure that the message reaches everyone, regardless of their preferred communication style.

Encourage open dialogue and feedback: Invite questions, comments, and suggestions from your team and organization. This promotes understanding and buy-in and allows you to gather valuable insights and ideas for refining the vision.

Share stories and real-world examples: Use case studies, anecdotes, and success stories to illustrate the relevance and importance of the product vision in action. This can make the vision more relatable and engaging for your team.

Connect the Vision to Goals and Objectives

To make the product vision tangible for your team, it's important to translate it into specific goals and objectives. This helps ensure everyone understands how their daily work contributes to achieving the vision. Here are some steps to connect the vision to goals and objectives:

Break down the vision into smaller goals: Identify the key milestones and outcomes required to realize the product vision. These milestones will be the foundation for your team's goals and objectives.

Set SMART goals and objectives: Ensure these goals and objectives are SMART (Specific, Measurable, Achievable, Relevant, and Time-bound). This provides a clear and actionable roadmap for your team to follow.

Assign ownership and accountability: Assign responsibility for each goal and objective to specific teams and individuals, ensuring everyone understands their role in achieving the vision.

Establish regular progress reviews: Schedule periodic reviews to assess progress towards the goals and objectives and adjust them to maintain alignment with the overall vision. This helps keep everyone focused and accountable.

Connect individual performance to the vision: Incorporate the goals and objectives related to the product vision into individual performance evaluations and career development discussions. This reinforces the importance of the vision and encourages team members to prioritize it in their daily work.

By following these steps, you can effectively connect your product vision to your team's goals and objectives, providing a clear and actionable roadmap for achieving the vision and ensuring everyone understands their role in its success.

Create a Visual Representation of the Vision

Visual representations can be powerful tools to convey complex ideas and concepts simply and engagingly. Creating a visual representation of your product vision can help your team and organization grasp the essence of the vision more effectively. Here's how you can make a compelling visual representation:

Choose the proper format: Depending on the nature of your product and the complexity of the vision, you might opt for a simple diagram, an infographic, a storyboard, or even a short video. Choose the format that best communicates the essence of your vision.

Keep it simple and focused: Avoid cluttering your visual representation with too much information or detail. Focus on the key elements of your vision, and ensure that the visual is easy to understand at a glance.

Use clear and consistent visual language: Stick to a consistent visual style and use clear, concise text to explain the critical aspects of your vision. This will help ensure that your visual representation is easily digestible and memorable.

Make it shareable and accessible: Ensure that all team members and stakeholders can easily share and access your visual representation. This might include creating digital versions for sharing via email or internal communication platforms and printed copies for display in communal areas.

Iterate and refine: Seek feedback from your team and stakeholders on the effectiveness of your visual representation, and make adjustments as necessary to ensure that it communicates your product vision.

Engage and Empower Your Team

To truly bring your product vision to life, you need the support and dedication of your team. Engaging and empowering your team to contribute to the vision fosters a sense of ownership and promotes a shared commitment to its success. Here's how to engage and empower your team in the pursuit of your product vision:

Involve your team in the vision creation process: Seek input and ideas from your team members as you develop and refine your product vision. This will help them feel more connected to the vision and invested in its success.

Encourage open communication: Foster a culture of open communication and collaboration where team members feel comfortable sharing their ideas, concerns, and feedback about the product vision.

Provide ongoing support and resources: Equip your team with the tools, training, and resources to help them contribute effectively to realizing the product vision. This includes providing clear guidance on expectations, priorities, and decision-making processes.

Celebrate successes and learn from setbacks: Acknowledge and celebrate the achievements and progress made by your team towards the product vision. At the same time, encourage a learning mindset and use setbacks as opportunities to reflect, learn, and grow.

Be a champion for the vision: As a product manager, it's essential to strongly advocate for the product vision within your team and across the organization. Demonstrate your commitment to the vision through your actions, communication, and decision-making, and inspire your team to do the same.

By engaging and empowering your team to pursue your product vision, you can foster a sense of shared ownership and commitment that will drive your team's passion, creativity, and determination to achieve the vision.

Align Your Stakeholders

Achieving your product vision often requires the support and collaboration of various stakeholders, such as executives, other departments, partners, and investors. Ensuring these stakeholders align with your product vision is crucial to its success. Here's how to align your stakeholders with your product vision:

Identify key stakeholders: Start by identifying the key stakeholders who directly or indirectly influence the success of your product vision. This may include senior executives, team leaders, marketing and sales teams, customer support teams, and external partners or investors.

Communicate the vision effectively: Share your product vision with your stakeholders clearly, concisely, and compellingly. Use the visual representation and other communication tools you've developed to ensure they understand the vision and its significance.

Highlight the benefits for stakeholders: Clearly articulate the benefits each stakeholder group will gain from successfully realizing the product vision. This may include increased revenue, improved customer satisfaction, or enhanced brand reputation.

Engage stakeholders in the process: Seek input and feedback from your stakeholders, and involve them in relevant aspects of the product development process. This will help them feel more connected to the vision and invested in its success.

Establish regular communication channels: Set up common communication channels to keep stakeholders informed of progress towards the product vision, challenges encountered, and opportunities for collaboration. This can include periodic reports, meetings, presentations, and more informal updates via email or internal communication platforms.

Be responsive to stakeholder concerns: Address any concerns or questions stakeholders raise promptly and transparently. Demonstrate your commitment to the product vision by responding to feedback and working collaboratively to overcome obstacles and challenges.

By aligning your stakeholders with your product vision, you can create a supportive ecosystem that enables your team to successfully deliver on the vision and achieve the desired outcomes for your product and organization.

In conclusion, communicating and evangelizing your product vision is essential to your role as a product manager. By clearly articulating your vision, creating a compelling visual representation, engaging and empowering your team, and aligning your stakeholders, you can inspire passion, creativity, and commitment to successfully realizing your product vision. Remember that your vision serves as the guiding light for your product journey, and it's up to you to keep that light shining brightly for all to see.

Key Takeaways

- Alignment around the product vision ensures unity, efficient decision-making, and a higher likelihood of success.
- Effective communication, connection of the vision to goals, visual representation, team engagement, and stakeholder alignment are key steps.
- Product managers should articulate the vision, engage and empower the team, and align stakeholders to realize the product vision.

Chapter 5

Evolution of Product Vision: The Necessity of Change

"It is not the strongest of the species that survives, nor the most intelligent; it is the one most adaptable to change." -Charles Darwin

A static product vision can significantly hinder your product's success. Here's why it's essential to be open to adapting and evolving your vision:

Market changes: Technological advancements, new competitors, and shifting customer preferences can quickly transform the market landscape. An adaptive product vision can help you stay ahead of these changes and remain competitive.

Product maturity: As your product matures and gains traction, its goals, and priorities may change. An evolving product vision can help you maintain focus and direction as your product grows.

Organizational shifts: Your company's goals, strategies, and resources may change over time, impacting your product vision. Being open to adapting your vision ensures alignment with the broader organizational context.

Learning from experience: As you develop and launch your product, you'll gain valuable insights and feedback from your customers, team, and stakeholders. These learnings can inform the evolution of your product vision, ensuring it remains relevant and effective.

Your product vision must be adaptable and flexible in today's fast-paced world. It should evolve and grow alongside your product, team, and market. This section will dive deeper into developing your product vision and provide examples of successful and unsuccessful adaptations.

Market Changes

Technological advancements, new competitors, and shifting customer preferences can quickly transform the market landscape. An adaptive product vision can help you stay ahead of these changes and remain competitive.

Successful example: Apple's vision has evolved from focusing on personal computers to a broader vision of delivering consumer electronics and services. This evolution allowed Apple to expand into new markets, such as smartphones and wearables, and maintain its position as a market leader.

Unsuccessful example: Kodak, once a dominant player in the photography industry, failed to adapt its vision to the digital age. The company's vision remained focused on traditional film photography, causing it to miss the shift to digital cameras, ultimately leading to its downfall.

Product Maturity

As your product matures and gains traction, its goals, and priorities may change. An evolving product vision can help you maintain focus and direction as your product grows.

Successful example: Amazon started as an online bookstore, but its vision evolved to become the world's most customer-centric company, encompassing retail, technology, and services. This expansion allowed Amazon to diversify its offerings and become a global powerhouse.

Unsuccessful example: Yahoo, once a pioneer in internet search and email services, struggled to evolve its product vision as the market changed. The company failed to adapt to the rise of mobile, social media, and search competitors like Google, resulting in a loss of market share and relevance.

Organizational Shifts

Your company's goals, strategies, and resources may change over time, impacting your product vision. Being open to adapting your vision ensures alignment with the broader organizational context.

Successful example: Microsoft's vision shifted from "a computer on every desk and in every home" to "empower every person and every organization on the planet to achieve more." This change allowed Microsoft to embrace cloud computing, mobile devices, and other emerging technologies, revitalizing the company's growth and innovation.

Unsuccessful example: BlackBerry's vision centered on secure and efficient communication devices for business professionals. However, the company failed to adapt its vision to the broader consumer market, ultimately losing out to more versatile smartphones offered by Apple and Android.

Learning from Experience

As you develop and launch your product, you'll gain valuable insights and feedback from your customers, team, and stakeholders. These learnings can inform the evolution of your product vision, ensuring it remains relevant and effective.

Successful example: Slack started as an internal communication tool for a game development company but evolved its vision to become a team collaboration and productivity platform. The company learned from user feedback and experiences, enabling it to pivot and create a product that resonated with a broader audience.

Unsuccessful example: MySpace was an early social media giant but failed to evolve its product vision in response to user feedback and changing preferences. As a result, MySpace could not compete with the rise of Facebook and other social media platforms, eventually fading into obscurity.

In summary, evolving your product vision is crucial for navigating the ever-changing technological, business, and customer expectations landscape. By being open to change and learning from your experiences, you can ensure your product vision remains a powerful guiding force that drives your product's success and growth.

When and How to Revisit Your Product Vision

Regularly revisiting and reevaluating your product vision is critical to ensuring it remains relevant and effective. Here's when and how to rethink your product vision:

Establish a review cadence: Set a regular schedule for reviewing and reassessing your product vision, such as every quarter or after major product milestones. This ensures that your vision remains top-of-mind and enables you to make proactive adjustments.

Gather input and feedback: Solicit input and feedback from your team, customers, and stakeholders as part of your vision review process. This helps you identify gaps, opportunities, or areas for improvement in your vision.

Assess alignment with goals and objectives: Evaluate how well your product vision aligns with your product and organizational goals and objectives. If you notice misalignments or gaps, consider updating your vision to reflect your priorities and desired outcomes.

Consider external factors: Examine the market landscape, technological trends, and competitive dynamics to identify any changes or disruptions that could impact your product vision. Adapt your vision as needed to ensure it remains relevant and compelling in the face of these external factors.

Iterate and refine: Based on your review and assessment, update and refine your product vision as needed. Communicate any changes to your team and stakeholders, and update your visual representation and other communication tools to reflect the corrected vision.

Prioritize Strategically

When making changes to your product vision, it's essential to prioritize updates that are most relevant and impactful to your product's success. Avoid making frequent, minor changes that can create confusion or a sense of instability. In this section, we'll delve into the importance of strategic prioritization and provide examples of effective and ineffective prioritization when updating a product vision.

Focus on High-Impact Changes

Not all changes to your product vision are created equal. Focus on updates that address market shifts, customer needs, or new opportunities rather than making insignificant tweaks that impact your product's direction.

Successful example: Netflix shifted its vision from being the best global DVD rental service to becoming the best global streaming service. This high-impact change allowed Netflix to capitalize on the emerging market for streaming content and maintain its leadership position in the face of competition from new entrants like Amazon Prime Video and Hulu.

Unsuccessful example: A software company frequently changes its product vision to reflect minor updates in technology trends or buzzwords without considering the broader implications for its product strategy and user experience. This lack of strategic focus leads to confusion among the team and stakeholders and dilutes the impact of the product vision.

Maintain Stability and Consistency

While it's essential to be open to change, it's equally crucial to maintain a sense of continuity and stability in your product vision. Prioritize changes that align with your product's core values and long-term goals, and avoid making updates that conflict with your established vision.

Successful example: Google's product vision has evolved to encompass a broader range of technologies and services. However, its core mission to "organize the world's information and make it universally accessible and useful" has remained consistent. This stability has allowed Google to expand into new areas like cloud computing and artificial intelligence while maintaining its focus on delivering value to users.

Unsuccessful example: A start-up that constantly changes its product vision to chase the latest market trends or investor interests may need help establishing a solid brand identity and loyal customer base. This lack of consistency can make it difficult for the team to stay aligned and focused on long-term goals, ultimately hindering the product's success.

Balance Adaptability and Stability

Strategic prioritization involves balancing adaptability and maintaining stability in your product vision. Be open to change when necessary and impactful, but don't allow your vision to become so malleable that it loses its guiding power.

Successful example: Adobe transitioned its vision from being the leader in desktop publishing software to providing creative tools and services in the cloud. This change allowed Adobe to embrace the shift to cloud-based software while maintaining its core focus on empowering creatives to bring their ideas to life.

Unsuccessful example: A mobile app company frequently changes its product vision to target different demographics or market segments without considering the long-term implications for its product strategy, user experience, or brand identity. This lack of strategic focus leads to a disjointed product and user experience, making it difficult for the company to attract and retain a loyal user base.

By prioritizing strategically when updating your product vision, you can ensure that your vision remains a powerful guiding force for your product's success while remaining responsive to the ever-changing landscape of technology and business. This approach helps you maintain focus, alignment, and stability as your product evolves and grows.

Involve Your Team in the Process

Involving your team in the process is crucial when updating your product vision. Collaboration and open communication help ensure your team remains engaged, aligned, and committed to the vision. This section will explore how to effectively involve your team in updating your product vision and discuss the benefits of fostering a collaborative and inclusive environment.

Encourage Open Dialogue

Open communication is vital to involve your team in updating your product vision. Encourage team members to share their perspectives, insights, and concerns. This allows you to gather diverse viewpoints and ensures that the corrected vision considers your team's collective wisdom.

Host brainstorming sessions: Organize brainstorming sessions where team members can freely discuss ideas, identify potential issues, and suggest improvements to the existing product vision. This collaborative approach can lead to innovative solutions and help ensure the updated vision aligns with the team's expertise and experience.

Solicit feedback: Actively seek feedback from team members at different stages of the updating process. This can help identify any confusion or misalignment and allows you to address concerns before they become more prominent.

Foster a Sense of Ownership

Involving your team in updating your product vision helps foster a sense of ownership and commitment to the product's success. When team members feel their input is valued and their contributions matter, they're more likely to be motivated and engaged in the product's development.

Delegate responsibilities: Assign team members specific responsibilities related to updating the product vision, such as researching market trends, analyzing customer feedback, or facilitating brainstorming sessions. This helps each team member feel a sense of ownership and responsibility for the vision's success.

Celebrate contributions: Recognize and celebrate the contributions of team members throughout the updating process. This helps reinforce the importance of collaboration and the value of each team member's input.

Maintain Transparency and Trust

Transparency and trust are essential for creating a collaborative and inclusive environment where team members feel comfortable sharing their thoughts and ideas. Be open and honest about the reasons behind updating the product vision, the goals you hope to achieve, and the challenges you anticipate.

Share updates regularly: Keep your team informed about the progress and outcomes of the updating process. Regular updates help maintain transparency and ensure team members feel connected to the evolution of the product vision.

Address concerns proactively: Be proactive in addressing any problems or questions that arise during the updating process. This helps build trust and demonstrates your commitment to an open and inclusive environment.

By involving your team in updating your product vision, you can create a more robust and well-rounded vision that benefits from diverse perspectives and expertise. This collaborative approach helps ensure your team remains engaged, aligned, and committed to the product's success as the vision evolves and adapts to changing circumstances.

Key Takeaways

- The product vision should adapt to market changes, product maturity, organizational shifts, and learnings from experience.
- Regular review and reevaluation of the product vision are vital.
- Changes to the product vision should focus on high-impact changes.
- It's essential to balance stability and consistency in the product vision while being open to changes.
- The team should be involved in the process of updating the product vision

Chapter 6

Expressing Your Vision: Engaging Stakeholders

"The single biggest problem in communication is the illusion that it has taken place." - George Bernard Shaw

A compelling product vision guides your team and helps get buy-in from stakeholders, such as investors, partners, and customers. This chapter will discuss effectively communicating your product vision to these stakeholders, ensuring they understand and support your vision and ultimately contribute to your product's success.

Tailor Your Message to Your Audience

One of the most important aspects of communicating your product vision is understanding each stakeholder group's unique perspectives, priorities, and interests. By tailoring your message to resonate with each specific audience, you'll be more successful in creating buy-in and generating support for your vision. Here, we'll delve deeper into how you can craft your communication for internal and external stakeholders.

Investors: These stakeholders are primarily interested in the market opportunity, growth potential, and how your product vision addresses a significant need or pain point. When communicating with investors, highlight your product's unique value proposition and competitive advantage. Demonstrate how your vision aligns with market trends and customer demands, and provide a clear view of the expected return on investment.

Customers: For customers, the focus should be on the benefits and value your product brings to their lives. Emphasize your commitment to customer success, and showcase how your product vision translates into tangible features and improvements that matter to them. Explain how your vision addresses their specific needs or problems, and use relatable examples and customer testimonials to illustrate the impact of your product.

Partners: When communicating with partners, showcase the synergies between your product and their offerings and explain how your vision aligns with their goals and objectives. Highlight your product's compatibility with their existing solutions, and emphasize the value of working together to achieve common objectives. Discuss the potential for collaboration and mutual growth, and provide examples of how the partnership can result in shared success.

Internal Stakeholders (Team Members, Executives, etc.): Your internal stakeholders play a vital role in executing your product vision and achieving success. Communicate how the vision connects to the company's mission, strategy, and values to ensure alignment and commitment. Explain each department and team member's role in realizing the vision, emphasizing the importance of collaboration and cross-functional support.

Team Members: Focus on how the product vision inspires and guides their daily work, and provide clear expectations and goals tied to the vision. Foster a sense of ownership and pride in contributing to the product's success, and encourage open dialogue to ensure everyone understands and supports the vision.

Executives: Demonstrate how your product vision aligns with the company's broader strategic objectives and contributes to overall business success. Provide a high-level overview of the product roadmap, and discuss the resources and support required to execute the vision. Be prepared to address any concerns or risks and present a clear plan for overcoming potential challenges.

By understanding the priorities and interests of each stakeholder group and tailoring your message accordingly, you'll be more effective in generating buy-in and support for your product vision. This ensures everyone is on the same page and contributes to your product's long-term success.

Utilize Different Communication Channels

To ensure that your product vision reaches all stakeholders and resonates with them, it's crucial to leverage various communication channels. Each channel has unique strengths and weaknesses, and by using a mix of them, you can maximize the impact of your message. Let's explore the most effective communication channels for sharing your product vision and the best ways to utilize them.

Presentations and meetings: In-person or virtual presentations and meetings are a powerful way to communicate your product vision to stakeholders. These settings allow for real-time interactions, questions, and feedback, which can help you gauge how well your message is received and address any concerns. Create engaging, visually appealing presentations that support your message, and practice your delivery to ensure a smooth and impactful experience.

Emails and newsletters: Emails and newsletters are an excellent way to keep stakeholders informed about the progress and updates related to your product vision. Use a clear subject line and concise, informative content to grab attention and convey your message effectively. Include relevant visuals, links, and resources that provide additional context and support your vision.

Internal communication platforms: Utilize internal communication platforms like Slack, Microsoft Teams, or intranets to share updates, discuss ideas, and gather feedback on your product vision. These platforms encourage collaboration and open dialogue, allowing your team members to feel more connected and engaged with the vision.

Social media and blogs: Leverage your company's channels and blogs to share your product vision with external stakeholders like customers, partners, and investors. Use a mix of engaging content formats like articles, videos, infographics, and podcasts to reach a wider audience and make your message more shareable. Monitor comments and interactions to gauge the response and address questions or concerns.

One-on-one conversations: Sometimes, the most effective way to communicate your product vision is through one-on-one conversations. These interactions allow for more in-depth, personalized discussions and can help you build strong relationships with key stakeholders. Be prepared to listen, ask questions, and adapt your message based on the feedback you receive.

Utilizing a mix of communication channels ensures that your product vision reaches all stakeholders and has the desired impact. Remember to tailor your message and approach for each channel, and be prepared to engage in open and constructive conversations to strengthen buy-in and support for your vision.

Use Powerful Storytelling Techniques

To effectively communicate your product vision to both internal and external stakeholders, it's essential to use powerful storytelling techniques. Stories can engage, inspire, and resonate with people personally, making them more likely to embrace your vision. Here are some tips for incorporating storytelling into your communication:

Be authentic: Share true stories that reflect the core values and principles behind your product vision. Authenticity helps build trust and credibility with your audience, making them more likely to connect with your message.

- When Airbnb struggled in its early days, the founders shared their personal stories of renting out their apartments to make ends meet, showcasing the origins and purpose behind their platform. This genuine story helped establish trust and credibility with their audience.

Appeal to emotions: People are more likely to remember and be influenced by stories that evoke emotions. Use descriptive language, anecdotes, and personal experiences to create an emotional connection with your audience.

- Apple's famous "Think Different" campaign focused on celebrating the rebels and visionaries who changed the world, evoking their audience's feelings of inspiration and aspiration and connecting them with Apple's vision to create innovative products.

Focus on the customer: Frame your product vision in terms of the value it brings to customers. Share stories of how your product will solve real problems and improve people's lives, making it easier for your audience to understand and relate to your vision.

- Tesla shares stories of how their electric cars are helping to reduce carbon emissions and contribute to a more sustainable future, emphasizing their products' positive impact on the environment and customers' lives.

Use relatable examples: Use examples and analogies your audience can easily understand and relate to. This helps to make your product vision more tangible and accessible, allowing stakeholders to see the real-world impact it can have.

- In its early marketing, Dropbox used a simple analogy: "Your files, anywhere." This helped convey the idea that their service made it easy to access files from any device at any time.

Keep it concise and engaging: While providing enough context and detail is essential, keeping your storytelling brief and interesting is equally crucial. Avoid overwhelming your audience with too much information, and focus on delivering a straightforward, compelling narrative.

- Google's mission statement, "To organize the world's information and make it universally accessible and useful," is a clear, concise, and engaging expression of their product vision that resonates with their audience.

Be Consistent and Clear

When communicating your product vision, it's crucial to be consistent and transparent in your messaging. This helps avoid confusion, reinforces your vision's importance, and builds credibility with stakeholders. Here are some tips for ensuring consistency and clarity in your communication:

Use simple, straightforward language: Avoid jargon, buzzwords, and overly complex language that can make your message difficult to understand. Instead, opt for simple, clear language that is accessible to a broad range of stakeholders.

- When explaining the functionality of their search engine, Google often states, "Just type what you're looking for and hit 'Enter,'" keeping the language simple and accessible.

Ensure alignment across all channels: Ensure your product vision is communicated consistently across all channels, including internal meetings, presentations, marketing materials, and social media. This helps to create a unified message and reinforces the importance of your vision.

- Slack's communication across its website, social media, and other channels consistently focuses on its vision to "make work simpler, more pleasant, and more productive."

Provide clear context and rationale: When sharing your product vision, provide clear context and rationale for why it matters. Explain how it aligns with your company's overall strategy and goals and the value it brings to customers.

- In its mission to "unlock the potential of human creativity," Spotify consistently communicates the value its platform brings to both artists and listeners, making the rationale for its product vision clear.

Establish a consistent visual identity: A solid visual identity can help to reinforce your product vision and make it more memorable. Develop a consistent visual style for your communication materials, such as presentations, infographics, and marketing collateral, that reflects the essence of your product vision.

- Coca-Cola's consistent use of red and white colors and its iconic script logo create a strong visual identity that reinforces its brand and product vision.

Regularly review and update your messaging: As your product evolves and your company grows, it's essential to periodically review and update your messaging to ensure it remains consistent and relevant. This helps to maintain clarity and alignment among stakeholders and keeps your product vision at the forefront of everyone's mind.

- As Amazon expanded beyond online retail, they updated their messaging to reflect their broader vision of being "Earth's most customer-centric company," ensuring that their communication remained relevant as their business evolved.

By incorporating powerful storytelling techniques and ensuring consistency and clarity in your communication, you can effectively share your product vision with internal and external stakeholders, driving engagement, enthusiasm, and success.

Key Takeaways

- Tailoring communication to each audience, using a variety of communication channels, and employing storytelling techniques are crucial.
- Consistency and clarity in communication should be maintained.
- Regular review and updates of the messaging as the product and company evolve is important.

Chapter 7

Reflection: Achieving Mastery of Your Product Vision

"Vision is the art of seeing what is invisible to others." - Jonathan Swift

Congratulations on completing the introduction to product vision of your product management journey! You now understand the importance of a well-defined product vision and its crucial role in guiding your product's development and success.

Summarizing the previous chapters dedicated to product vision, we've explored:

- **The Essence of a Product Vision:** A strong product vision serves as the foundation of your product, providing inspiration, direction, focus, alignment, and differentiation. It is the guiding force that keeps your team and stakeholders on the same page and moving toward a common goal.
- **Crafting Your Product Vision Statement:** We dis-

cussed the elements that make a great product vision statement—user-centric, clear & concise, inspiring, long-term, and unique & memorable. These characteristics help to ensure your vision is both practical and compelling.

- **Aligning Your Team and Organization Around the Vision:** Communicating your vision internally is crucial for aligning your team, stakeholders, and the entire organization around a shared purpose. We covered various methods to achieve alignment, such as workshops, vision boards, and team presentations.

- **Evolving Your Product Vision Over Time:** As your product and market grow, so should your product vision. We explored techniques for adapting your vision to changing circumstances, like regular vision reviews, incorporating customer feedback, and staying informed about market trends.

- **Communicating Your Vision to Stakeholders:** To secure buy-in and support, it's essential to share your vision with internal and external stakeholders effectively. We examined powerful storytelling techniques, the importance of consistency and clarity, and using various communication channels to share your vision.

As we wrap up this chapter, please take a moment to reflect on the insights and examples we've shared. Consider how you can apply these lessons to your product management practice. Remember, a compelling product vision is the cornerstone of your product's success, and mastering this aspect of product management will set you on a path toward achieving your goals.

In the next chapter, we'll delve deeper into the world of product management, exploring product strategy and providing you with tools and techniques to make informed decisions and drive your product's growth. So, stay tuned and get ready to elevate your product management skills even further!

Chapter 8

Setting Your Strategic Compass: Success with Product Strategy

"Good business leaders create a vision, articulate the vision, passionately own the vision, and relentlessly drive it to completion."
-Jack Welch

If product vision is the lighthouse guiding your ship, product strategy is the map that helps you navigate the waters to reach your destination. It's the plan you follow to turn your vision into reality, the foundation upon which your product's success is built. A well-defined product strategy helps you chart a course through the choppy waters of a competitive market and ensures that you are sailing in the right direction. In this chapter, we'll explore the essentials of crafting a winning product strategy and how it contributes to the overall success of your product.

Imagine you're a captain of a ship embarking on a long journey. You have a clear destination(your product vision), but to get there, you need a detailed map (your product strategy). This map will help you make critical decisions, like which route to take, what resources you'll need, and how to react to unexpected obstacles along the way. Without a strategy, your ship could quickly veer off course or become lost in the stormy seas of product development.

So, let's embark on this exciting journey and learn how to create a robust product strategy. We'll explore the importance of understanding your target market and customer needs, analyzing the competitive landscape, crafting your unique value proposition, prioritizing initiatives and allocating resources, setting strategic goals and objectives, and leveraging technology and innovation in your product strategy.

Understanding Your Target Market and Customer Needs

Before building a product, you must know who you're building it for and what problems you're trying to solve. You can't create a successful outcome without first understanding your target market and the specific needs of your customers. In this section, we'll dive deep into the importance of customer research, defining your target market segments, and uncovering your customers' needs, wants, and pain points.

Analyzing the Competitive Landscape

No product exists in a vacuum, and understanding the competitive landscape is crucial for crafting an effective product strategy. You need to be aware of the players in your market, their strengths and weaknesses, and the opportunities and threats they present. We'll explore various techniques to analyze your competition, identify market gaps, and uncover opportunities for differentiation.

Crafting Your Unique Value Proposition

Now that you know your target market and understand the competitive landscape, it's time to define what sets your product apart. Your unique value proposition (UVP) is the secret sauce that differentiates your product from the competition and conveys its specific benefits to your customers. We'll discuss crafting a compelling UVP that resonates with your target audience and drives product adoption.

Prioritizing Initiatives and Resource Allocation

Developing a product requires making countless decisions, from determining which features to build to allocating resources among various initiatives. In this section, we'll discuss prioritizing your ambitions and allocating resources effectively to maximize the impact on your product strategy. We'll cover multiple frameworks and techniques to help you make informed decisions and ensure your efforts align with your overall product goals.

Setting Strategic Goals and Objectives

Goals and objectives are the milestones that guide your product development efforts and ensure that you stay on track. They provide a clear direction for your team and help you measure progress toward your product vision. In this section, we'll delve into setting clear and measurable goals and objectives that align with your product strategy, ensuring everyone on your team is working towards a common purpose.

Leveraging Technology and Innovation in Your Product Strategy

In today's fast-paced, technology-driven world, staying ahead of the competition often means embracing innovation and leveraging cutting-edge technology. In this section, we'll explore how to incorporate technology and innovation into your product strategy, ensuring your product remains relevant, competitive, and future-proof. We'll discuss the role of emerging technologies, staying up-to-date with industry trends, and fostering a culture of innovation within your product team.

With this introduction, we've set the stage for a deep dive into the world of product strategy. In the following sections, you'll learn the steps and best practices to create a winning strategy to help you turn your product vision into reality. We'll cover everything from understanding your customers and competition to crafting a compelling value proposition, prioritizing initiatives, setting goals, and leveraging technology and innovation.

So, buckle up and get ready to explore the exciting realm of product strategy! With these essential tools and insights, you'll be well-equipped to chart a course that leads your product to success and sets it apart in the competitive landscape. Happy sailing!

Chapter 9

Deciphering Your Market: The Key to Customer Needs

"Get closer than ever to your customers. So close that you tell them what they need well before they realize it themselves." - Steve Jobs

This section will explore the importance of understanding your target market and customer needs and how this knowledge is crucial in shaping your product strategy. We'll discuss techniques for market segmentation, creating buyer personas, conducting customer research, and prioritizing customer needs.

Market Segmentation: Finding Your Tribe

Before addressing your customers' needs, you must know who they are. Market segmentation divides potential customers into groups based on shared characteristics, such as demographics, behaviors, and needs. You can tailor your product strategy to meet their unique requirements by identifying and focusing on specific segments.

Demographic Segmentation: Group customers based on age, gender, income, education, or occupation. For example, a fitness app might target millennial women with a college education and a higher-than-average income. Conversely, a poor example would be if the same app targeted people of all ages and incomes without considering their specific needs and preferences.

Geographic Segmentation: Divide customers by location, such as country, region, or city. A local food delivery service could focus on urban areas with a high concentration of restaurants and busy professionals. A poor example would be a business that tries to cater to the entire world without adapting its offerings to local tastes and preferences.

Psychographic Segmentation: Classify customers based on lifestyle, personality traits, values, and interests. A sustainable fashion brand might focus on environmentally conscious consumers who value ethical and eco-friendly products. In contrast, a poor example would be targeting all fashion enthusiasts without considering their specific values and interests.

Behavioral Segmentation: Categorize customers by their purchasing habits, product usage, and loyalty. A subscription box service might target frequent online shoppers who enjoy trying new products. A poor example would be targeting all consumers, regardless of their shopping habits and brand loyalty.

Creating Buyer Personas: Putting a Face on Your Customers

Buyer personas are fictional representations of your ideal customers based on actual data and market research. Creating detailed buyer personas can help your product team understand your target audience's motivations, challenges, and preferences.

To create effective buyer personas, consider the following:

Demographics: Age, gender, income, location, etc. For example, "Tech-savvy Tom" is a 28-year-old software engineer living in San Francisco, earning $100,000 annually. A poor example would be a vague description like "young professionals," which doesn't provide specific information about the target audience.

Goals and Motivations: What are your customers trying to achieve? What drives their decisions? "Eco-friendly Emily" might be motivated by reducing her carbon footprint and supporting ethical brands. A poor example would be a persona with generic motivations like "wants to save money," which doesn't provide actionable insights for your product strategy.

Pain Points: What problems do your customers face? How can your product solve them? "Busy Ben," a working parent, might struggle to prepare healthy meals for his family. A poor example would be a persona with pain points unrelated to your product, such as "doesn't like traffic," which doesn't help you understand how your product can address their needs.

Preferred Channels: How do your customers prefer to communicate and interact with brands? "Social Media Sally" might prefer engaging with brands on Instagram and Twitter, while "Email Enthusiast Eric" prefers newsletters and promotional emails. A poor example would be assuming all customers use the same channels, such as only focusing on social media without considering other communication preferences.

Conducting Customer Research: Listen, Learn, and Adapt

To truly understand your customers, you must gather data and insights about their needs, preferences, and behaviors. There are various methods to conduct customer research, including:

Surveys: Collect quantitative data from a large sample of customers through online or in-person questionnaires. Be sure to ask specific questions related to your product and target audience. A poor example would be using generic survey templates that don't provide actionable insights.

Interviews: Conduct one-on-one interviews with customers to gather in-depth qualitative insights. Focus on understanding their motivations, pain points, and preferences. A poor example would be asking leading questions that influence the customer's response, resulting in biased data.

Focus Groups: Bring together a small group of customers to discuss their needs, opinions, and experiences related to your product. Ensure the conversation is moderated effectively to encourage open and honest feedback. A poor example would be holding focus groups without a clear agenda or not recording the session for future analysis.

Observational Research: Observe customers interacting with your product or service in person or through tools like website analytics. This can help you identify usability issues, user behaviors, and preferences. A poor example would be relying solely on your observations without gathering data from other sources.

Prioritizing Customer Needs: Focus on What Matters Most

Once you've gathered insights about your target market and customer needs, prioritizing them is essential. This will help you create a product strategy that addresses the most critical issues and delivers the most significant value to your customers. Consider using techniques like the Kano Model or the MoSCoW method to prioritize customer needs based on their impact and feasibility.

A good example of prioritizing customer needs would be focusing on the features with the most significant potential to solve your customers' pain points and differentiate your product from competitors. A poor example would be trying to address all customer needs simultaneously, resulting in a bloated product with little focus or competitive advantage.

This section explored the importance of understanding your target market and customer needs and how this knowledge can shape your product strategy. By segmenting your market, creating buyer personas, conducting customer research, and prioritizing customer needs, you can develop a product strategy that truly resonates with your audience and drives success.

Key Takeaways

- Understanding your customers' shared characteristics allows for a more focused product strategy. Segmentation can be based on demographics, geography, psychographics, and behaviors.
- Constructing fictional representations of your ideal customers using real data and market research is helpful for understanding and defining your target demographic. It includes considerations of demographics, goals and motivations, pain points, and preferred communication channels.
- Gathering data about your customers' needs, preferences, and behaviors is crucial. Research methods include surveys, interviews, focus groups, and observational research. Each method should be applied thoughtfully for actionable insights.
- Once you understand your customers, prioritizing their needs helps you create a strategy that addresses key issues and delivers significant value. Techniques such as the Kano Model or the MoSCoW method are helpful for prioritization.

Chapter 10

Unveiling Allies and Adversaries: Dissecting the Competitive Landscape

"If you know the enemy and know yourself, you need not fear the result of a hundred battles." - Sun Tzu

This chapter will explore the importance of understanding the competitive landscape and how this knowledge can help shape your product strategy. By assessing your competitors' strengths and weaknesses and identifying potential opportunities and threats, you can create a product strategy that sets you apart and gives you a competitive edge.

Identifying Your Competitors: Look Beyond the Obvious

The competitive analysis involves studying your competitors to identify their strengths and weaknesses, helping you to position your product more effectively in the market. To conduct a competitive analysis, you can:

Identify your main competitors: List the top companies offering similar products or services in your market. This should include direct competitors (those offering the same type of product) and indirect competitors (those offering a different product that solves the same customer need).

Analyze their product offerings: Evaluate your competitors' products in terms of features, benefits, and pricing. This can help you understand their value proposition and identify potential gaps in the market that your product can fill.

Assess their market positioning: Understand how your competitors position themselves, including their target audience, messaging, and branding. This can give you insights into their strategy and help you differentiate your product.

For example, a new meal-kit delivery service might analyze the offerings of established competitors, such as Blue Apron and HelloFresh. These competitors primarily target busy professionals and families, offering a limited range of vegetarian options. Seeing an opportunity, the new service could focus on providing a wider variety of plant-based meals to cater to a growing market of health-conscious consumers. A poor example would be trying to compete directly with established brands with a clear differentiation or unique value proposition.

Identify Opportunities for Differentiation

Once you've conducted a competitive analysis, you can use the insights gained to identify opportunities for differentiation. Differentiation refers to the unique aspects of your product or service that set it apart from competitors, making it more appealing to your target market. To identify opportunities for differentiation, consider the following:

Unique features or benefits: Identify aspects of your product that competitors do not offer or that you can provide in a superior way. This could include exclusive functionalities, excellent performance, or a better user experience.

Niche markets: Explore niche markets or underserved segments of your target audience that may not be well-served by existing competitors. Focusing on a specific niche can help you build a loyal customer base and stand out in a crowded market.

Branding and messaging: Develop a unique brand identity and messaging that resonates with your target audience and sets you apart from competitors. This could include a distinctive visual style, tone of voice, or brand values.

For example, a new ride-sharing app might differentiate itself from established competitors like Uber and Lyft by focusing on eco-friendly transportation and offering a fleet of exclusively electric vehicles. This would appeal to environmentally conscious consumers seeking a more sustainable transportation option. A poor example would be launching a ride-sharing app with no clear differentiation, making it difficult for users to see any reason to choose it over the established competitors.

A SWOT (strengths, weaknesses, opportunities, threats) analysis can be a helpful tool in this process.

Strengths: What do your competitors do well? What advantages do they have over your product or company? For example, a competitor may have a strong brand reputation, extensive distribution network, or innovative technology.

Weaknesses: Where do your competitors fall short? What challenges do they face that you can exploit? Examples could include a lack of product customization options, outdated technology, or poor customer service.

Opportunities: Can you address any gaps in the market or unmet customer needs? This might include new technologies, trends, or underserved customer segments that your competitors have overlooked.

Threats: What external factors or trends could pose risks to your business? These might include new market competitors, changing customer preferences, or regulatory changes.

Keep an Eye on Emerging Trends and Competitors

Staying informed about emerging trends and potential new competitors is essential for maintaining a competitive edge in your market. This allows you to adapt your product strategy as needed, ensuring you stay ahead of the competition and capitalize on new opportunities. Here's how you can stay on top of the latest trends and competitors:

Monitor industry news and reports: Regularly read industry publications, subscribe to newsletters, attend conferences, and follow thought leaders in your field to stay informed about the latest trends, technologies, and competitors. Set up news alerts for specific keywords related to your industry and make it a habit to dedicate time each week to stay updated. This will help you anticipate changes in the market and adjust your product strategy accordingly.

Conduct ongoing competitive analysis: Update your competitive analysis regularly, such as quarterly or bi-annually, to keep track of new entrants, product updates from existing competitors, and any changes in market dynamics. Use tools like SWOT analysis, feature comparisons, and pricing analysis to evaluate your competitors' strengths and weaknesses. This will help you stay agile and quickly adapt your product strategy when needed.

Engage with your target audience: Talk to your customers, potential customers, and industry influencers through interviews, surveys, or social media interactions to get insights into their needs, preferences, and pain points. Monitor online forums and communities where your target audience discusses their challenges and experiences. This will help you identify trends and potential opportunities for differentiation.

For example, a project management software company could notice the rise of remote work (an emerging trend) and the growing demand for tools that facilitate collaboration across distributed teams. By staying informed about this trend, the company could pivot its product strategy to add features that cater to remote groups, such as enhanced communication tools, task delegation, and progress tracking. This would allow them to stay ahead of competitors who have yet to adapt to this trend.

A poor example would be a company that fails to recognize the importance of mobile platforms, leading them to focus exclusively on desktop applications. At the same time, competitors successfully capture the growing mobile market.

Be Prepared to Pivot Your Product Strategy

The market constantly changes, and your product strategy should be flexible enough to adapt. Be prepared to pivot your product strategy if new opportunities, threats, or trends emerge that could impact your product's success. Here's how you can ensure your product strategy remains adaptable:

Establish a culture of agility and adaptability: Encourage a mindset of continuous improvement and learning within your team. Foster a culture where experimentation and learning from failures are embraced. This will help you quickly adapt to changes in the market and identify new opportunities or threats.

Set regular strategy review meetings: Schedule periodic meetings, such as monthly or quarterly, to review your product strategy, assess its effectiveness, and make necessary adjustments. Invite key stakeholders to these meetings, so everyone stays aligned and can contribute their perspectives. This will help you remain proactive and ensure your product aligns with your business goals.

Remain customer-focused: Always prioritize your customer's needs, as their preferences and pain points may change over time. Regularly engaging with your target audience through feedback sessions, user testing, and surveys will help you stay informed about their evolving needs and adapt your product strategy accordingly.

For example, a company that develops a social media management tool may notice a decline in the popularity of a specific social media platform and the rise of a new one. By pivoting their product strategy to support the new platform, they can maintain their competitive edge and continue to cater to the changing needs of their customers.

A poor example would be a company that ignores the emergence of new competitors with innovative features and sticks to its original product strategy, eventually losing market share and falling behind.

Understanding the competitive landscape is crucial to developing an effective product strategy. By analyzing your competitors' strengths and weaknesses, staying informed about emerging trends and new competitors, and being prepared to pivot your product strategy when needed, you can ensure that your product remains competitive and continues to deliver value to your customers. Regularly updating your competitive analysis and engaging with your target audience will help you stay agile and quickly adapt to changes in the market, ultimately contributing to the success of your product.

Leveraging Competitive Intelligence to Inform Your Product Strategy: A Deeper Look at Benchmarking

Benchmarking compares your product or business processes against your competitors or industry leaders' best practices and performance standards. It helps you understand your position in the market, identify areas for improvement, and learn from the successes of others. Here's a more detailed look at how you can use benchmarking in your product strategy:

Feature benchmarking:

- Identify the key features that customers value in your product category. Create a comprehensive list that covers all aspects of the customer experience.
- Assess your product against your competitors' products in terms of these features. Rate each feature on a scale, considering factors such as functionality, usability, and innovation.
- Highlight the areas where your product excels and those where it lags. This helps you to prioritize product improvements and focus on the most impactful features.

Example: A good practice is to use a spider chart or radar chart to visualize the feature benchmarking results, making it easy to spot areas of strength and weakness. A poor practice would be focusing only on a few key features, ignoring other important aspects of the customer experience.

Performance benchmarking:

- Determine the key performance indicators (KPIs) most relevant to your product and customers. These may include metrics like customer satisfaction, user engagement, and conversion rates.

- Gather data on your product's performance and that of your competitors. Use reliable sources like market research reports, customer reviews, and industry publications.

- Compare your product's performance against the industry average and top performers. Identify areas where you are outperforming or underperforming your competitors.

Example: A good practice is to track your performance over time, allowing you to monitor trends and measure the impact of product improvements. A poor approach would be to rely on a single data point or snapshot in time, ignoring the bigger picture.

Price benchmarking:

- Research the pricing strategies of your competitors. Take note of factors such as price tiers, discounts, and promotions.

- Evaluate your product's pricing relative to your competitors. Consider your product's perceived value, features, and target market.

- Adjust your pricing strategy as needed to remain competitive while ensuring profitability. Use the insights gained from benchmarking to inform your pricing decisions.

Example: A good practice is to conduct regular price benchmarking exercises to stay up-to-date with market trends and competitor moves. A poor approach would be to set your pricing based solely on cost or gut feeling without considering the competitive landscape.

By conducting a thorough benchmarking analysis, you can make more informed decisions regarding your product strategy and set realistic goals for improvement. Additionally, it can help you identify new opportunities, spot industry trends, and ensure that your product remains competitive in the market.

Key Takeaways

- Identify and analyze your direct and indirect competitors, their product offerings, and market positioning to guide your product's strategic positioning.
- Use insights from your competitive analysis to identify unique features, niche markets, and branding opportunities that set your product apart.
- Stay informed about emerging trends and potential competitors by monitoring industry news, conducting regular competitive analyses, and engaging with your target audience.
- Foster agility, conduct regular strategy reviews, and remain customer-focused to ensure your product strategy adapts to changing market conditions.
- Benchmarking against competitors or industry leaders in features, performance, and pricing can guide your product strategy, set realistic goals, and ensure competitiveness.

Chapter 11

Forging Your Unique Offering: The Value Proposition

"Don't find customers for your products; find products for your customers." - Seth Godin

Your unique value proposition (UVP) is a clear, concise statement that communicates the primary benefit your product offers to customers and sets it apart from competitors. A strong UVP is essential for a successful product strategy, as it helps you attract and retain customers, focus your marketing efforts, and guide product development. In this section, we'll explore the process of crafting a compelling UVP for your product.

Understand Your Customer's Pain Points and Desires

To craft an effective UVP, you must first understand your customer's pain points and desires. This in-depth knowledge enables you to focus on the aspects of your product that matter most to your target audience.

Conduct comprehensive market research: You must conduct thorough market research before identifying your customers' pain points and desires. This involves exploring your target market's demographics, behaviors, and preferences to gain valuable insights. Utilize tools such as Google Analytics, industry reports, and social media analytics to gather quantitative data.

Utilize qualitative research methods: To delve deeper into your customers' needs and emotions, employ qualitative research methods such as interviews, focus groups, and ethnographic studies. These methods enable you to engage directly with your customers, allowing you to collect richer data and better understand their motivations, frustrations, and aspirations.

Analyze customer feedback and reviews: Examine existing feedback and reviews to identify common themes and recurring issues. You can gather customer feedback from social media, online review platforms, and customer support interactions. This information can be a goldmine for uncovering pain points and desires that your product can address.

Map customer journeys: Creating journey maps can help you visualize your customers' experiences, from initial awareness to purchase and beyond. These maps can reveal potential pain points and unmet desires at each customer journey stage, enabling you to address them with your product.

Example: A good practice is to create customer personas representing different segments of your target audience, helping you empathize with their needs and preferences. A poor approach would be to make assumptions about your customers without validating them through research.

Define the Core Benefits Your Product Offers

Once you understand your customer's pain points and desires, you can identify your product's core benefits. These benefits should directly address your customers' needs, setting your product apart from competitors.

List your product's features and benefits: Start by listing your product's features and the benefits they provide to customers. Consider how each feature solves a pain point or satisfies a desire. Remember, customers are more interested in your product's benefits than the features themselves.

Identify your product's unique differentiators: Determine what sets your product apart from competitors. Focus on the aspects of your product that are truly unique and valuable to customers. These differentiators could include superior functionality, better user experience, or more competitive pricing.

Prioritize the most compelling benefits: Not all benefits are equally important to your customers. Based on your research and analysis, prioritize the most relevant benefits to your target audience. These high-priority benefits should form the foundation of your UVP.

Involve cross-functional teams: To gain a comprehensive understanding of your product's benefits, involve cross-functional teams, such as product, marketing, and sales, in identifying and prioritizing them. These diverse perspectives can help you uncover benefits you may have overlooked and ensure your UVP resonates with customers.

Example: A good practice is using a prioritization matrix or other decision-making tool to evaluate and rank each benefit's importance objectively. A poor approach would be focusing solely on features without considering the benefits they provide customers.

Articulate Your UVP Clearly and Concisely

Once you've identified your product's core benefits, it's essential to articulate your UVP clearly and concisely. A well-crafted UVP quickly communicates your product's unique value, making it easy for customers to understand why they should choose your product over competitors.

Keep it simple: Your UVP should be easy to understand and free of jargon or technical terms that might confuse potential customers. Aim to express your UVP in a single, straightforward sentence that communicates the essence of your product's unique value.

Be specific: Generic or vague statements won't make your product stand out from the competition. Instead, be clear about your product's benefits and the problems it solves for customers. Use concrete language and data to support your claims when possible.

Focus on emotional appeal: People often make decisions based on emotions rather than logic, so your UVP should evoke an emotional response from potential customers. Highlight the positive feelings your customers will experience by using your product, such as relief, excitement, or satisfaction.

Test and refine your UVP: Your UVP may not be perfect on the first attempt. Gather feedback from customers, colleagues, and stakeholders to identify areas for improvement. This feedback will refine your UVP until it effectively communicates your product's unique value.

Example: A good practice is to use A/B testing to evaluate the effectiveness of different UVP variations with your target audience. A poor approach would be to use a long, convoluted sentence that is difficult for customers to understand and remember.

Integrate Your UVP into Marketing & Sales Materials

To ensure your UVP is consistently communicated to potential customers, it's crucial to integrate it into your marketing and sales materials. This includes incorporating your UVP into various touchpoints along the customer journey, from initial awareness to purchase and beyond.

Update your website and landing pages: Ensure your UVP is prominently displayed on your homepage and relevant landing pages. This helps potential customers understand your product's unique value when they visit your site.

Incorporate your UVP into marketing materials: Integrate your UVP into marketing materials such as brochures, email campaigns, social media posts, and advertisements. This consistent messaging helps reinforce your product's unique value in the minds of potential customers.

Train your sales team: Ensure it is well-versed in your UVP and can communicate it effectively to potential customers. Provide sales representatives with talking points, scripts, and other resources to help them articulate your product's unique value during sales conversations.

Monitor and measure your UVP's impact: Regularly evaluate your UVP's effectiveness in driving customer interest and conversion. Use tools such as web analytics, customer surveys, and sales data to assess the impact of your UVP on your business's performance. Refine your UVP and its implementation as needed based on this data.

Example: A good practice is to create a style guide outlining how your UVP should be presented across different channels and materials, ensuring a consistent look and feel. A poor approach would be to neglect to incorporate your UVP into key customer touchpoints, leading to inconsistent messaging and missed opportunities.

Adapt Your UVP for Different Audiences

Adapting your unique value proposition (UVP) for different audiences is crucial to effectively communicate your product's value. Customers from diverse backgrounds and needs may perceive your product's value differently. By tailoring your UVP, you can address each customer segment's unique needs and preferences.

Define your customer segments: Start by identifying the various customer segments your product serves. This will help you recognize their different needs and expectations. Consider demographics, psychographics, and behavior to create distinct profiles for each segment.

Research each segment's needs: For each customer segment, dive deeper into their specific problems and desired benefits. Conduct market research, surveys, interviews, or focus groups to gather insights to inform a more targeted UVP for each audience.

Develop audience-specific UVPs: Craft variations of your UVP that highlight the unique aspects and benefits most relevant to each customer segment. Ensure the core message of your product's exceptional value remains consistent across all tailored UVPs.

Test and refine tailored UVPs: Measure the effectiveness of your tailored UVPs on customer engagement, conversion, and retention. Use this data to refine your UVPs for each audience segment and optimize your marketing strategy.

Example: A good practice is to use A/B testing to gauge the effectiveness of different UVP variations for specific customer segments. A poor approach would be to use a generic, one-size-fits-all UVP that may not resonate with particular audiences.

Continuously Review and Update Your UVP

Your product, customers, and market conditions are subject to change, so reviewing and updating your UVP regularly ensures it remains relevant and compelling.

Track changes in customer need: Monitor trends, customer feedback, and industry developments to stay informed of shifting customer preferences and pain points. Adjust your UVP to reflect evolving customer needs.

Stay up-to-date with market developments: Keep an eye on new competitors, emerging technologies, and other market changes that could impact your product's UVP. Adapt your UVP to stay ahead and maintain your competitive edge.

Incorporate product improvements into your UVP: As you enhance your product and add new features, evaluate how these changes affect your UVP. Update your UVP to reflect your product's current value and benefits, ensuring it remains appealing and accurate.

Communicate UVP updates internally and externally: Make sure your team is aware of any updates to your UVP, and incorporate the changes into their marketing and sales efforts. Update your marketing materials and customer touchpoints to reflect the revised UVP.

Example: A good practice is to hold regular team meetings or workshops to discuss your product's UVP, assess its continued relevance, and make any necessary updates. A poor approach would be to create a UVP and never revisit it, even as your product and market evolve.

Key Takeaways

- Conduct both quantitative and qualitative market research.
- Utilize customer feedback, reviews, and journey mapping.
- Identify and prioritize product benefits that solve customer needs.
- Collaborate with cross-functional teams to highlight unique differentiators.
- Create a simple, specific UVP with emotional appeal.
- Continually test and refine your UVP.
- Display UVP prominently across marketing materials and websites.
- Train the sales team on UVP and monitor its impact.
- Understand customer segments and tailor UVPs to each segment's needs.
- Keep UVP updated with customer needs, market trends, and product improvements.

Chapter 12

Sorting Your Priorities: Initiatives and Resource Management

"Efficiency is doing things right; effectiveness is doing the right things." - Peter Drucker

Effective product strategy involves making tough decisions on which initiatives to prioritize and how to allocate resources. In this section, we'll explore techniques for evaluating and prioritizing initiatives and ensuring the optimal allocation of resources to maximize impact.

Establish Evaluation Criteria

Before you can prioritize initiatives, it's essential to establish a clear set of evaluation criteria. These criteria will help you assess each initiative's value, feasibility, and impact.

Strategic alignment: Evaluate how well the initiative aligns with your product strategy and business goals. Prioritize initiatives that contribute directly to your strategic objectives.

Customer impact: Determine the potential impact of each initiative on your customers. Consider both the magnitude of the benefit and the number of customers who will be affected.

ROI: Estimate the return on investment (ROI) for each initiative. Compare the expected benefits to the costs and resources required to implement the initiative. Prioritize those with the highest ROI.

Feasibility: Assess the technical and operational feasibility of each initiative. Consider the complexity, required expertise, and potential risks.

Time to market: Estimate the time to deliver each initiative. Prioritize those that can be completed quickly to capitalize on market opportunities or address pressing customer needs.

Apply Prioritization Frameworks

Several prioritization frameworks can help you objectively rank initiatives based on your evaluation criteria. Choose a framework that aligns with your product strategy and organization's decision-making style.

MoSCoW: The MoSCoW framework categorizes initiatives into four groups: *Must-have*, *Should-have*, *Could-have*, and *Won't-have*. This method provides a clear overview of priorities and helps teams focus on the most critical initiatives.

RICE: The RICE framework scores initiatives based on *Reach*, *Impact*, *Confidence*, and *Effort*. Calculate a RICE score for each initiative and rank them accordingly.

Kano Model: The Kano Model classifies features based on their potential impact on customer satisfaction. Identify features that will delight customers, satisfy their basic needs, or prevent dissatisfaction, and prioritize accordingly.

Value vs. Effort: Create a simple two-dimensional matrix with value on the vertical axis and effort on the horizontal axis. Plot each initiative on the matrix and prioritize those with the highest value and lowest effort.

Allocate Resources Wisely

Once you've prioritized your initiatives, allocate resources to ensure the successful implementation of each project.

Balance short-term and long-term initiatives: Allocate resources to a mix of short-term, high-impact initiatives and long-term, strategic projects to maintain momentum and achieve lasting success.

Consider team capabilities and capacity: When allocating resources, assess your team's expertise, workload, and capacity. Ensure that the right people work on suitable projects to maximize effectiveness.

Monitor and adjust resource allocation: Continuously monitor project progress and resource utilization. Be prepared to adjust resource allocation as needed to address changing priorities, new opportunities, or unexpected challenges.

Communicate resource allocation decisions: Transparently communicate your resource allocation decisions to your team and stakeholders. This will help ensure buy-in and support for your prioritization choices.

Continuously Evaluate and Adjust Priorities

Your product strategy and priorities will evolve over time. Regularly review your initiatives and resource allocation to ensure they remain aligned with your strategic goals and market conditions.

Hold periodic prioritization sessions: Schedule regular meetings or workshops to reassess your priorities and resource allocation. Involve key stakeholders to ensure their input and buy-in.

Be agile and adaptable: Stay responsive to changes in customer needs, market conditions, and competitive dynamics. Be prepared to adjust your priorities and resource allocation as needed to stay ahead of the curve.

Learn from past experiences: Reflect on the outcomes of completed initiatives and use this information to inform your future prioritization decisions. Identify what worked well, what didn't, and what can be improved.

Foster a culture of continuous improvement: Encourage your team to embrace change and continuously seek ways to optimize your product strategy, prioritization, and resource allocation. This will help your organization stay nimble and adaptable in a rapidly changing market.

Balance Stakeholder Interests

Managing the interests of various stakeholders is a crucial aspect of prioritizing initiatives and allocating resources. Keep open communication channels with stakeholders and strive to find a balance that satisfies their needs and aligns with your product strategy.

Identify critical stakeholders: Understand who your key stakeholders are and what their interests and concerns are. These may include customers, investors, partners, and internal teams.

Engage stakeholders in the decision-making process: Involve stakeholders in the prioritization and resource allocation to build consensus and foster collaboration.

Manage expectations: Communicate your priorities and resource allocation decisions to stakeholders, explaining the rationale behind your choices. This will help manage expectations and minimize potential conflicts.

Be prepared to negotiate and compromise: Recognize that different stakeholders may have conflicting interests and priorities. Be prepared to negotiate and find compromises that balance these interests while aligning with your product strategy.

In conclusion, prioritizing initiatives and allocating resources is critical to product strategy. By establishing clear evaluation criteria, applying prioritization frameworks, allocating resources wisely, and continuously evaluating and adjusting your priorities, you can ensure that your product strategy remains aligned with your business goals and delivers maximum value to your customers and organization.

Key Takeaways

- Align initiatives with product strategy and business goals.
- Evaluate customer impact and ROI.
- Assess feasibility and time to market.
- Use frameworks like MoSCoW, RICE, Kano Model, or Value vs. Effort.
- Balance short-term and long-term initiatives.
- Consider team capabilities and capacity.
- Monitor resource utilization and adjust allocation as needed.
- Communicate resource allocation decisions clearly.
- Hold regular prioritization sessions.
- Be agile and adaptable to market changes.
- Learn from past experiences and foster continuous improvement.
- Identify and engage key stakeholders in decision-making.

- Manage stakeholder expectations and be ready to negotiate and compromise.

Chapter 13

Establishing Your Strategic Milestones: Goals and Objectives

"Setting goals is the first step in turning the invisible into the visible."
- Tony Robbins

Setting strategic goals and objectives is essential to a successful product strategy. These goals and objectives provide a clear direction for your product and help you measure progress and success over time. This section will discuss the importance of setting goals and objectives, tips for creating effective goals, and how to align them with your overall product strategy.

The Importance of Setting Goals and Objectives

Clear direction: Goals and objectives give your team a sense of direction and purpose, ensuring everyone is working towards the same result. This alignment prevents confusion, streamlines decision-making, and reduces the likelihood of conflicting priorities. By setting specific targets, your team can develop a shared understanding of success and focus their efforts on achieving these outcomes.

Measure progress and success: Goals and objectives provide a framework for measuring progress, which helps you determine if your product strategy is on track and identify areas that require improvement. Regularly reviewing your progress against your goals enables you to make data-driven decisions and adjust your strategy when necessary. With clear benchmarks to evaluate your progress, it's easier to ascertain if your efforts yield the desired results.

Align team members and stakeholders: Clearly defined goals and objectives help align the efforts of team members and stakeholders, ensuring everyone works together to achieve the desired outcomes. This alignment is essential for maintaining a collaborative environment and fostering a sense of shared ownership and responsibility. When everyone is on the same page, it's easier to coordinate activities, communicate effectively, and make decisions that support the overall product strategy.

Drive decision-making: Goals and objectives guide decision-making, helping you prioritize initiatives and allocate resources effectively. Setting targets creates a clear framework for evaluating different initiatives' potential impact and value. This enables you to make informed decisions about which projects to pursue, which to postpone, and which to abandon altogether. Your goals and objectives can also help you determine how to allocate resources, such as budget, personnel, and time, to maximize the chances of success.

Foster accountability and motivation: Setting goals and objectives encourages team members to take ownership of their work and strive for excellence. When team members understand the objectives they're working towards, they're more likely to feel motivated and accountable for their performance. This sense of responsibility can drive them to find creative solutions, persevere through challenges, and celebrate their achievements as they progress toward their goals.

Facilitate communication and transparency: Establishing clear goals and objectives enables better communication within your team and with external stakeholders. Sharing your targets and progress updates creates a sense of transparency that builds trust and credibility. This open communication fosters collaboration and can help you uncover valuable insights, feedback, and ideas to inform your product strategy and drive continuous improvement.

In summary, setting strategic goals and objectives plays a crucial role in your product strategy by providing direction, measuring progress, aligning team members and stakeholders, driving decision-making, fostering accountability and motivation, and facilitating communication and transparency.

Creating SMART Goals and Objectives

Developing effective goals and objectives is crucial for your product strategy's success. Utilizing the SMART framework ensures that your goals are Specific, Measurable, Achievable, Relevant, and Time-bound. This section will delve deeper into each element of SMART goals, providing a comprehensive understanding and practical examples to illustrate their significance.

Specific

A specific goal or objective clearly communicates the desired outcome. It leaves no room for ambiguity or confusion. Specific goals answer the essential questions of who, what, where, when, and why. For example, instead of stating, "Increase user engagement," a specific goal could be, "Increase the average time spent on our app by 20% for users in the 18-24 age group within six months." You give your team a clear direction by specifying the target audience, desired improvement, and timeframe.

To make your goals more specific, you can use the following tips:

- Use clear and concise language to describe the outcome.
- Break down complex goals into smaller, more manage-

able tasks.

- Identify the key stakeholders involved and their responsibilities.

Measurable

Measurable goals and objectives enable you to track progress and assess the success of your efforts. By incorporating quantifiable metrics, such as numbers, percentages, or timeframes, you create a tangible way to evaluate performance. For instance, "Improve customer satisfaction" could become measurable by stating, "Increase our customer satisfaction rating from 80% to 90% within the next year." Including a metric and a deadline allows you to monitor progress and make data-driven decisions.

Some tips for making your goals measurable include:

- Identify relevant key performance indicators (KPIs) that align with your objectives.
- Establish a baseline for each metric and set targets for improvement.
- Regularly review progress and adjust your approach as needed.

Achievable

Considering your current resources, constraints, and market conditions, achievable goals and objectives are realistic and within reach. Overly ambitious targets can demoralize your team and result in disappointment when expectations aren't met. Conversely, goals that are too easy may not encourage innovation or improvement. For example, instead of aiming to "Double our market share within a month," a more achievable goal might be "Increase our market share by 10% within the next year."

To ensure your goals are achievable, consider the following:

- Assess your team's capabilities and resources, including time, budget, and expertise.
- Evaluate potential obstacles and develop strategies to overcome them.
- Regularly revisit and adjust your goals based on the changing circumstances.

Relevant

Relevant goals and objectives align with your overall product strategy and address the needs of your target audience. They contribute to the achievement of your broader vision and mission. For instance, if your product's primary focus is to provide a seamless user experience, a relevant goal could be, "Reduce the number of reported bugs by 50% within the next three months."

To create relevant goals, keep these tips in mind:

- Review your product vision and strategy to ensure alignment.
- Consider your target market's needs and preferences.
- Align your goals with the priorities and values of your organization.

Time-bound

Time-bound goals and objectives have a definitive deadline, instilling a sense of urgency and helping your team prioritize their efforts. A time-bound goal specifies when the objective should be achieved, whether within a week, a month, a quarter, or a year. For example, "Increase monthly active users" could become time-bound by stating, "Increase monthly active users by 30% within the next six months."

To set time-bound goals, follow these guidelines:

- Establish realistic deadlines based on the scope and complexity of each goal.
- Consider any dependencies or external factors that may impact the timeline.

- Monitor progress regularly and adjust deadlines if necessary while maintaining a sense of urgency.

Incorporating the SMART framework into your goal-setting process will create a clear, focused, and actionable roadmap for your team and organization. This approach ensures that your strategic goals and objectives align with your product vision, address relevant customer needs, and can be effectively monitored and adjusted as needed.

In summary, developing SMART goals and objectives for your product strategy entails:

- Creating specific, well-defined targets that answer the essential questions of who, what, where, when, and why.

- Establishing measurable, quantifiable metrics to track progress and evaluate success.

- Consider your resources, constraints, and market conditions to ensure your goals are achievable.

- Aligning your objectives with your overall product vision and strategy to ensure relevance.

- Setting time-bound deadlines to encourage prioritization and maintain a sense of urgency.

By following these guidelines, you can establish a solid foundation for your product strategy, providing a clear direction for your team and increasing the likelihood of achieving your desired outcomes.

Aligning Your Team and Stakeholders Around Your Goals and Objectives

Achieving your strategic goals and objectives requires setting them and ensuring your team and stakeholders are aligned and committed to working towards them. This alignment is crucial for fostering collaboration, streamlining decision-making, and maximizing the chances of success. In this section, we'll explore how to align your team and stakeholders effectively around your goals and objectives.

Transparent communication: To get everyone on board, it's essential to communicate your goals and objectives clearly, highlighting their purpose, expected outcomes, and deadlines. Use various communication channels such as meetings, presentations, and written documents to ensure everyone understands the goals and their role in achieving them.

For example, instead of vaguely stating that you want to "increase revenue," specify that you aim to "increase revenue by 20% in the next 12 months by expanding into new markets and improving our pricing strategy."

Link goals to the broader vision and strategy: Help your team and stakeholders see how the goals and objectives contribute to the overall product vision and strategy. This connection enables them to appreciate the value of their efforts and keeps them motivated to work towards their goals.

For instance, explain how a goal like "reducing customer support response time by 50%" aligns with the broader vision of "providing an exceptional customer experience that sets us apart from the competition."

Collaborative goal-setting: Engage your team and stakeholders in the goal-setting process by soliciting their input, ideas, and concerns. This collaborative approach fosters a sense of ownership and commitment to the goals, as everyone feels their voice has been heard and their perspective is considered.

For example, hold brainstorming sessions or workshops where team members can share their thoughts on the proposed goals and suggest possible improvements or alternatives.

Monitor and share progress: Regularly track and share progress towards the goals and objectives with your team and stakeholders. This transparency keeps everyone informed, motivated, and accountable, allowing for timely course corrections.

Consider using tools like dashboards or reports that visually display progress, making it easy for everyone to see how they're doing and where they need to focus their efforts.

Recognize achievements and learn from setbacks: Celebrate milestones and accomplishments as a team, acknowledging individual and collective efforts. This recognition boosts morale and encourages continued commitment to the goals. When setbacks occur, analyze the reasons, learn from them, and adjust your approach as needed.

For example, hold team celebrations or give shout-outs to team members who have contributed significantly. When things don't go as planned, hold constructive discussions to understand what went wrong and how to improve moving forward.

By aligning your team and stakeholders around your goals and objectives, you create a focused, engaged, and collaborative environment that increases the chances of achieving your strategic targets and ultimately realizing your product vision.

Key Takeaways

- Strategic goals and objectives are vital for product strategy, providing direction, measuring progress, aligning teams, and facilitating communication.
- Goals should be crafted following the SMART framework: Specific, Measurable, Achievable, Relevant, and Time-bound.
- Transparent communication, goal relevance to the broader vision, collaborative goal-setting, and regular progress monitoring are essential for aligning the team and stakeholders around the goals.
- Recognizing achievements and learning from setbacks fosters a productive environment and helps ensure the success of the strategic targets.

Chapter 14

Reflection: Shaping Your Successful Product Strategy

"All men can see these tactics whereby I conquer, but what none can see is the strategy out of which victory is evolved." - Sun Tzu

You now understand the importance of a well-defined product strategy and its crucial role in developing and succeeding your product.

In this section on Product Strategy, we've explored:

Understanding Your Target Market and Customer Needs: By getting to know your target audience and identifying their pain points, you can create products and services that truly resonate with them.

Analyzing the Competitive Landscape: We discussed the importance of understanding your competition, conducting SWOT and benchmarking analyses, and staying informed about market trends to help inform your strategic decisions.

Crafting Your Unique Value Proposition: We covered developing a compelling value proposition that sets your product apart from competitors and communicates its value to customers.

Prioritizing Initiatives and Resource Allocation: We explored techniques for prioritizing projects and tasks, ensuring your team focuses on the most impactful work and efficiently uses resources.

Setting Strategic Goals and Objectives: We emphasized the importance of setting clear and measurable goals, aligning your team and stakeholders around these targets, and fostering collaboration and streamlined decision-making.

As we wrap up these chapters on product strategy, please take a moment to reflect on the insights and examples we've shared. Consider how you can apply these lessons to your product management practice. Remember, a well-crafted product strategy drives your product's growth and success. Mastering this aspect of product management will set you on a path toward achieving your goals.

In the following chapters, we'll dive into the world of product roadmaps, providing tools and techniques to create roadmaps that effectively communicate your product's vision, strategy, and priorities. So, stay tuned and get ready to elevate your product management skills even further!

Chapter 15

Guiding Your Journey: The Product Roadmap

"A goal without a plan is just a wish." - Antoine de Saint-Exupéry

Welcome to the next step of your product management journey! Having hoisted the sails of product vision and plotted the course with product strategy in the previous chapters, we're now ready to navigate the uncharted waters of product roadmaps. A product roadmap is an essential navigational tool that helps translate your product vision and strategy into actionable steps, guiding your crew and stakeholders through the product development voyage.

In this chapter, we'll explore the different aspects of product roadmaps, from their various types to the methods used to create, maintain, and communicate them effectively. As you read the following chapters on product roadmaps, you'll gain valuable insights and best practices for creating roadmaps that align with your product vision and strategy while providing a clear path forward for your product development efforts.

Why are product roadmaps essential? Roadmaps serve as a strategic chart that outlines the direction and timeline for your product's development. They help prioritize initiatives, set expectations for stakeholders, and ensure that everyone involved in the product development process works towards common goals. Roadmaps are critical for maintaining focus and providing a shared understanding of the product's objectives, making them indispensable in the world of product management.

As you progress through the chapters on product roadmaps, we'll discuss the various types of roadmaps, such as goal-oriented, feature-based, time-based, and theme-based. Each type serves a different purpose and suits different organizational contexts and needs. We'll help you determine which roadmap type best fits your organization, enabling you to create roadmaps that effectively communicate your product's direction.

Creating compelling product roadmaps is an art, much like charting the course for a successful voyage. We'll delve into the best practices for defining goals and objectives, prioritizing initiatives and features, and incorporating user feedback and data. You'll learn how to balance short-term and long-term goals, ensuring that your roadmap remains aligned with your product vision and strategy while addressing immediate needs and opportunities.

As products evolve and markets change, keeping your product roadmap up-to-date and relevant is essential, just as a seasoned captain adjusts the course in response to shifting winds and currents. We'll discuss the importance of regular reviews and updates and adapting your roadmap to changing market conditions and customer needs. Furthermore, we'll explore how to communicate roadmap changes to stakeholders, ensuring everyone remains aligned and informed throughout the product development process.

Visualizing and presenting your product roadmap is a crucial aspect of effective communication. We'll explore various visualization tools, techniques, and best practices for tailoring presentations to different stakeholders. You'll be better equipped to secure buy-in and support for your product initiatives by ensuring clarity and simplicity in your roadmap presentations.

Lastly, we'll address common roadmap challenges, such as managing competing priorities, addressing uncertainty and change, and ensuring stakeholder alignment and buy-in. By understanding these challenges and learning how to overcome them, you'll be better prepared to navigate the complexities of product roadmap management.

The following content will comprehensively understand product roadmaps, their importance, and the best practices for creating, maintaining, and communicating them effectively. Applying these insights to your product management practice, you'll be better equipped to guide your product development efforts and drive your product's success. So let's embark on this exciting journey and master the art of product road mapping!

Chapter 16

Diverse Paths: Understanding Different Product Roadmaps

"There are many paths to the top of the mountain, but the view is always the same." - Chinese proverb

Navigating the world of product management requires a keen understanding of the various types of product roadmaps that can be used to chart the course of your product's development. Much like a skilled sailor who selects the appropriate chart for the journey ahead, choosing the right type of roadmap for your product and organization is essential. The product roadmap serves as a strategic communication tool, visually representing the product's direction, highlighting the key milestones, and setting stakeholder expectations.

In this section, we'll explore five common types of product roadmaps: goal-oriented, feature-based, time-based, theme-based, and hybrid roadmaps. Each roadmap type has unique characteristics that cater to different product management styles, organizational structures, and industry demands. Understanding the nuances of each type will enable you to make informed decisions about the most suitable approach for your product's roadmap.

We'll discuss each type's unique characteristics, advantages, and use cases, helping you determine which roadmap best fits your needs. You'll learn how each roadmap type aligns with different organizational priorities and goals and how they facilitate effective communication with stakeholders.

Moreover, we'll also explore how the choice of the roadmap type can impact the product development process, team collaboration, and resource allocation. By examining the strengths and weaknesses of each roadmap type, you'll be better equipped to make informed decisions that lead to a successful product journey.

So, set your sails and prepare to embark on a voyage through the various product roadmap types, gaining the knowledge and insights needed to steer your product in the right direction and navigate the ever-changing waters of product management.

Goal-Oriented Roadmaps

A goal-oriented roadmap, also known as an outcome-based roadmap, focuses on your product's strategic objectives and desired outcomes rather than specific features or timelines. This type of roadmap is beneficial for organizations that prioritize impact and results over particular deliverables. Aligning the product roadmap with the organization's broader goals encourages a more flexible and adaptable approach to product development.

Advantages:

A clear focus on desired outcomes and impact: Goal-oriented roadmaps provide a high-level view of the product's strategic direction, emphasizing the organization's desired results. This helps to keep the team focused on the bigger picture and prevents them from getting lost in the details of specific features.

Encourages flexibility and adaptability: With a focus on goals and outcomes, goal-oriented roadmaps allow for greater flexibility in the product development process. As market conditions or customer needs change, the team can adjust its approach to achieve the desired outcomes without being locked into specific features or timelines.

Prioritization of initiatives: Goal-oriented roadmaps help teams prioritize initiatives based on their potential contribution to the objectives. This ensures that resources are allocated effectively and that the most impactful projects receive attention.

Use Cases:

Organizations with rapidly changing markets and customer need: In fast-paced industries, it's crucial to remain agile and adaptable. Goal-oriented roadmaps allow teams to respond quickly to changing market conditions and customer requirements, ensuring the product remains relevant and competitive.

Products in the early stages of development: During the initial stages of product development, priorities and features can change frequently as the team learns more about the market and customer needs. A goal-oriented roadmap provides the flexibility to navigate these changes and adapt the product's direction accordingly.

Companies with a strong focus on innovation and experimentation: For organizations prioritizing innovation and experimentation, a goal-oriented roadmap provides a framework for exploring new ideas and testing hypotheses while remaining focused on the desired outcomes.

Feature-Based Roadmaps

Feature-based roadmaps focus on the specific features and functionality that will be developed for your product. This type of roadmap is best suited for organizations with a clear understanding of their product's requirements and a stable market environment. By providing a detailed view of the planned features, a feature-based roadmap helps to set clear expectations for stakeholders and ensure that resources are allocated effectively.

Advantages:

Detailed overview of planned features: Feature-based roadmaps provide a comprehensive view of the product's development, outlining the specific features and functionality that will be delivered. This level of detail helps to set expectations for stakeholders and allows the team to plan and allocate resources effectively.

Clear expectations for stakeholders: By outlining the specific features that will be developed, a feature-based roadmap helps manage stakeholder expectations and ensure everyone is on the same page regarding the product's direction.

Better planning and resource allocation: With a clear view of the feature set, teams can plan their development efforts more effectively, allocating resources to the most critical features and ensuring deadlines are met.

Use Cases:

Mature products with a well-defined set of features and requirements: For products with a clear and stable set of requirements, a feature-based roadmap provides a straightforward and detailed plan for development. This helps to keep the team focused and ensures that resources are used effectively.

Organizations with a stable market environment and predictable customer need: In industries with relatively stable market conditions, a feature-based roadmap can ensure that the product remains competitive by providing a clear plan for addressing customer needs and expectations.

Teams that rely heavily on detailed planning and resource allocation: For groups that require a high level of planning and coordination, a feature-based roadmap provides the necessary structure and detail to ensure that development efforts are well-organized and resources are allocated effectively.

Time-Based Roadmaps

Time-based roadmaps outline your product's planned features and initiatives within specific time frames, such as quarters, months, or even weeks. This type of roadmap provides a clear view of the product's development timeline, helping teams to plan and allocate resources effectively. Time-based roadmaps are particularly useful for organizations with fixed deadlines or tight schedules.

Advantages:

A clear view of the development timeline: Time-based roadmaps provide a detailed picture of the product's development timeline, allowing teams to plan their work and allocate resources effectively. This helps to ensure that deadlines are met and that the product remains on track.

Better alignment with organizational goals and initiatives: By aligning the product's development timeline with the broader goals and initiatives of the organization, time-based roadmaps help to ensure that the product remains in sync with the company's overall strategy.

Improved communication and coordination: Time-based roadmaps help facilitate communication and coordination among team members, ensuring everyone knows the development schedule and their responsibilities.

Use Cases:

Products with fixed deadlines or tight schedules: For products with strict deadlines or tight development schedules, a time-based roadmap provides the necessary structure and detail to ensure that the product remains on track and that deadlines are met.

Organizations with strong alignment between product development and business goals: In companies where product development is closely aligned with business goals and initiatives, a time-based roadmap can help to ensure that the product remains in sync with the company's overall strategy.

Teams with a high level of interdependence and coordination: For teams that rely heavily on coordination and communication, a time-based roadmap can help to facilitate this by providing a clear view of the development timeline and individual responsibilities.

Hybrid Roadmaps

Hybrid roadmaps combine elements of goal-oriented, feature-based, and time-based roadmaps, providing a comprehensive and flexible view of the product's development. This type of roadmap is best suited for organizations with a mix of short-term and long-term objectives that require a balance between flexibility and detailed planning.

Advantages:

A comprehensive view of the product's development: Hybrid roadmaps provide a holistic view of the product's development, incorporating elements of goal-setting, feature planning, and timeline management. This comprehensive view helps to ensure that the product remains on track and that all aspects of development are considered.

Flexibility and adaptability: By combining elements of different roadmap types, hybrid roadmaps provide the flexibility to adapt to changing market conditions, and customer needs while maintaining a clear view of the product's overall direction.

Better alignment with organizational goals and initiatives: Hybrid roadmaps help to ensure that the product's development remains aligned with the broader goals and initiatives of the organization while still allowing for flexibility and adaptability in the development process.

Use Cases:

Organizations with a mix of short-term and long-term objectives: Hybrid roadmaps are well-suited to companies with a combination of short-term and long-term goals, as they balance detailed planning and flexibility.

Products with a mix of stable and evolving requirements: For products with a combination of regular and changing needs, a hybrid roadmap provides the necessary structure and adaptability to ensure that the product remains competitive and relevant.

Teams that balance detailed planning and flexibility: For groups that need careful planning and flexibility, a hybrid roadmap provides a comprehensive and adaptable framework for managing the product's development.

Theme-Based Roadmaps

Rather than specific features or timelines, theme-based roadmaps focus on high-level themes or strategic objectives that guide the product's development. This type of roadmap provides a more flexible approach to product planning, allowing teams to adapt to changing market conditions and customer needs while maintaining alignment with the product's overall direction.

Advantages:

Greater flexibility and adaptability: Theme-based roadmaps provide the flexibility needed to adapt to changing market conditions and customer needs while maintaining a clear view of the product's overall direction.

Better alignment with organizational goals and initiatives: By focusing on high-level themes and strategic objectives, theme-based roadmaps help to ensure that the product's development remains aligned with the broader goals and initiatives of the organization.

More accessible communication with stakeholders: Theme-based roadmaps can be easier to communicate to stakeholders, as they focus on high-level objectives rather than specific features or timelines. This can make it easier for stakeholders to understand and buy into the product's overall direction.

Use Cases:

Organizations with rapidly changing markets or customer needs: Theme-based roadmaps are well-suited to companies operating in rapidly changing markets or those with rapidly evolving customer needs, as they provide the necessary flexibility and adaptability to keep the product competitive and relevant.

Products with a solid strategic focus: For products with a strong strategic focus, a theme-based roadmap can help to ensure that the product's development remains aligned with the organization's broader goals and objectives.

Teams that prioritize adaptability over detailed planning: For groups that value adaptability and flexibility over careful planning, a theme-based roadmap provides a high-level framework for managing the product's development, allowing them to respond quickly to changing conditions.

Key Takeaways

Product roadmaps are strategic communication tools that visually represent product direction and setting stakeholder expectations.

Five common types of product roadmaps include goal-oriented, feature-based, time-based, hybrid, and theme-based roadmaps.

Goal-Oriented Roadmaps:

- Focus on strategic objectives and desired outcomes.
- Encourage flexibility and adaptability.
- Prioritize initiatives based on their potential contribution to the objectives.
- Ideal for organizations in rapidly changing markets, in early stages of development, or focusing on innovation.

Feature-Based Roadmaps:

- Focus on specific features and functionalities.
- Provide a detailed view of planned features.
- Help manage stakeholder expectations and facilitate

better planning and resource allocation.

- Best suited for mature products with well-defined requirements and organizations in stable market environments.

Time-Based Roadmaps:

- Outline product features and initiatives within specific time frames.
- Provide a clear view of the product's development timeline.
- Useful for organizations with fixed deadlines or tight schedules and those requiring strong alignment between product development and business goals.

Hybrid Roadmaps:

- Combine elements of goal-oriented, feature-based, and time-based roadmaps.
- Provide a comprehensive and flexible view of product development.
- Best for organizations with a mix of short-term and long-term objectives, or products with a mix of stable and evolving requirements.

Theme-Based Roadmaps:

- Focus on high-level themes or strategic objectives.
- Provide a flexible approach to product planning.
- Useful for organizations operating in rapidly changing markets, those with rapidly evolving customer needs, or products with a strong strategic focus.

Chapter 17

Structuring Your Path: Building an Effective Product Roadmap

"Failing to plan is planning to fail." - Alan Lakein

This section will discuss the key elements and best practices for creating an effective product roadmap. From establishing a clear vision and strategic objectives to prioritizing features and aligning stakeholders, these practices will help you navigate the complexities of product development and ensure your product stays on course.

Here are the main topics we'll cover in this section:

- Aligning the Roadmap with Product Vision and Strategy
- Identifying and Prioritizing Features and Goals
- Balancing Short-Term and Long-Term Priorities
- Collaborating with Cross-Functional Teams

- Communicating the Roadmap to Stakeholders
- Adapting the Roadmap to Changing Circumstances

By mastering the art of creating an effective product roadmap, you'll be better equipped to lead your product through the ever-evolving landscape of market demands, technological advancements, and stakeholder expectations. So, sharpen your axe, and let's dive into the world of product roadmaps.

Aligning the Roadmap with Product Vision and Strategy

The first step in creating an effective product roadmap is to ensure it aligns with your product vision and strategy. This alignment is essential because it helps maintain consistency and focus throughout the product development process, providing a clear direction for your team and stakeholders.

Here are some key aspects to consider when aligning your roadmap with your product vision and strategy:

Reflect the product vision: Your roadmap should be an extension of your product vision, encapsulating the same aspirational, long-term goals that inspire and motivate your team. Ensure the roadmap's objectives and milestones align with the overarching vision.

Example: If your product vision is to empower remote workers with seamless communication and collaboration tools, your roadmap should prioritize features and improvements that enhance remote collaboration and communication.

Stay true to the product strategy: Your product strategy outlines the high-level approach to achieving your vision. Your roadmap should act as a tactical plan to execute that strategy. Ensure the roadmap aligns with your strategic goals, unique value proposition, and target market needs.

Example: If your strategy involves differentiating your product through exceptional user experience, prioritize features and improvements that enhance usability, accessibility, and overall user satisfaction in your roadmap.

Keep strategic objectives in mind: Strategic objectives are the measurable outcomes that help you gauge the success of your product strategy. When crafting your roadmap, consider the objectives and how the planned features or initiatives contribute to achieving them

*Exam*ple: If your strategic objective is to increase user retention by 20% within the next year, prioritize features and improvements that address known pain points and drive user engagement.

By aligning your product roadmap with your product vision and strategy, you can create a coherent and focused plan that effectively guides your product's development journey.

Identifying and Prioritizing Features and Goals

Once your product roadmap is aligned with your vision and strategy, the next step is to identify and prioritize the features and goals that will help you achieve your desired outcomes. This process is critical, ensuring that your team focuses on the most impactful and valuable initiatives.

Here are some methods and best practices for identifying and prioritizing features and goals in your product roadmap:

Gather input from various sources: Collect feedback from multiple sources, such as user research, customer feedback, competitive analysis, and information from cross-functional teams, to identify potential features and goals.

Example: Conduct user interviews, analyze customer support tickets, and collaborate with sales and marketing teams to identify areas for improvement or new opportunities.

Utilize prioritization frameworks: Leverage established prioritization frameworks, such as the RICE (Reach, Impact, Confidence, Effort) model or the MoSCoW (Must-have, Should-have, Could-have, Won't-have) method, to objectively evaluate and rank the importance of each feature or goal.

Example: Using the RICE model, assign scores to each potential feature based on its reach, impact, confidence, and effort, and prioritize those with the highest overall score.

Balance customer needs with business goals: Ensure your prioritized features and plans address customer needs and your organization's strategic objectives. Strive to create a well-rounded roadmap that delivers value to your users while driving business growth.

Example: Prioritize features that solve vital customer pain points while contributing to your organization's revenue targets or market expansion goals.

Consider dependencies and constraints: When prioritizing features and goals, consider any dependencies or constraints, such as technical limitations, resource availability, or regulatory requirements, that may impact their implementation.

Example: If a high-priority feature requires significant architectural changes, you may need to adjust your roadmap to accommodate the necessary groundwork before focusing on that feature.

Be flexible and adaptable: As your product evolves and new information becomes available, be prepared to adjust your roadmap priorities accordingly. Stay responsive to changing customer needs, market trends, and competitor activities.

Example: If a new competitor emerges with a disruptive feature, you may need to reevaluate your priorities to respond effectively and maintain your competitive edge.

Communicate priorities clearly: Once you've determined the priorities for your product roadmap, ensure that they are communicated to your team and stakeholders. This transparency helps maintain alignment and manage expectations throughout the product development process.

Example: Regularly share your prioritized roadmap with your team and stakeholders, and provide context for the decisions made. Encourage open discussions and feedback to foster a collaborative environment.

By effectively identifying and prioritizing features and goals in your product roadmap, you'll allocate your resources efficiently and keep your team focused on the most impactful initiatives. This approach will help you deliver a product that meets your customers' needs and drives your organization toward its strategic objectives.

Collaborating with cross-functional teams to refine your roadmap

Involving cross-functional teams is essential to creating a genuinely effective product roadmap. This collaboration ensures that your roadmap reflects diverse perspectives and expertise, leading to better decision-making and a more robust, well-rounded product. Here are some tips for collaborating with cross-functional teams:

Hold regular roadmap review meetings: Meet key stakeholders from different departments, including development, design, marketing, and sales. Use these meetings to review and discuss the current state of the roadmap, gather feedback, and make any necessary adjustments.

Example: A development team might raise concerns about the feasibility of implementing a particular feature within the proposed timeline. This feedback can help you reassess your priorities and adjust the roadmap accordingly.

Encourage open communication: Create a culture that values open communication and feedback. Encourage team members to share their thoughts and ideas, even if they challenge the current roadmap.

Example: Set up a dedicated communication channel (e.g. , Slack, Microsoft Teams) where team members can share their thoughts and ideas about the roadmap and collaborate on potential improvements.

Leverage the expertise of each team: Each team brings unique insights and expertise to the table. Make sure to tap into their knowledge and experience when refining your roadmap.

Example: Consult with the marketing team to understand market trends and competitor activities, or collaborate with the design team to ensure your product's user experience aligns with your vision.

Iterate and refine: Continuously iterate on your roadmap based on the feedback and insights gathered from your cross-functional teams. This ongoing collaboration will help you create a more accurate and effective roadmap.

Example: After receiving feedback from multiple teams, you may realize that a particular feature needs to be deprioritized in favor of another, more critical initiative. Update your roadmap accordingly and communicate these changes to all stakeholders.

Select the Appropriate Roadmap Type

Selecting the correct roadmap type for your product and organization is crucial in effectively communicating your product's plan and progress. While several types of roadmaps are available, choosing one that best aligns with your product's needs, stage, and goals is essential. It's also important to consider the needs of your stakeholders when selecting the roadmap type, as different stakeholders may have different expectations and requirements.

Here are a few factors to consider when selecting the appropriate roadmap type:

Product Stage: The stage of your product's lifecycle may influence the type of roadmap you choose. For instance, a product in its early stages of development may benefit more from a goal-oriented or theme-based roadmap, focusing on high-level objectives and themes. In contrast, a mature product may require a feature-based or time-based roadmap detailing specific features and deadlines.

Stakeholder Needs: Different stakeholders may have varying preferences and requirements regarding the roadmap's format and level of detail. For example, executives might prefer a high-level roadmap focusing on strategic goals and objectives. At the same time, development teams may need a more detailed roadmap that outlines specific features and tasks. It's essential to strike a balance between the needs of different stakeholders and ensure your roadmap is easily accessible and understandable to everyone.

Organizational Culture: Your organization's culture and decision-making processes can also influence the choice of the roadmap type. If your organization values innovation and agility, a more flexible roadmap like a theme-based or goal-oriented one may be suitable. On the other hand, if your organization has a more structured approach to planning and execution, a time-based or feature-based roadmap may be more appropriate.

Adaptability: As product roadmaps often need to adapt to changes in the market, customer needs, and internal priorities, it's essential to choose a roadmap type that allows for flexibility and easy updates. Some roadmap types, like goal-oriented and theme-based roadmaps, are inherently more adaptable to change, while others, like time-based or feature-based roadmaps, may require more effort to update and maintain.

Remember that there's no one-size-fits-all roadmap; you may need to customize or combine different roadmap types to suit your needs. The key is to ensure that your chosen roadmap type effectively communicates your product's direction, priorities, and progress while being adaptable and relevant to your stakeholders' needs.

Balance Flexibility and Structure

Achieving the right balance between flexibility and structure in your product roadmap is vital to accommodate changing circumstances while maintaining the focus on your product's overarching goals and objectives. Striking this balance can be challenging, but it is essential for managing expectations, securing stakeholder buy-in, and fostering a productive and collaborative environment. Below, we'll delve into some strategies to help you balance flexibility and structure in your product roadmap:

Prioritize high-level objectives: Establish your product's strategic goals and objectives as the foundation of your roadmap. Focusing on high-level objectives enables your team to remain aligned with the big picture and facilitates adaptation when details of the roadmap change. For example, outline the desired outcome instead of listing specific features, such as improving user engagement or increasing conversion rates. This approach allows for flexibility in achieving the objectives while maintaining a clear direction.

Time horizons and progressive elaboration: Rather than assigning specific deadlines to every feature or task, consider using time horizons (e.g., short-term, medium-term, long-term) or progressive elaboration (providing more detail for near-term items while keeping long-term things more high-level). Time horizons allow your team to allocate resources and plan for the upcoming period. At the same time, progressive elaboration lets you revisit and refine long-term plans as more information becomes available.

Embrace change: Accept and plan for the inevitability of change in your roadmap. Schedule regular review and update cycles to ensure your roadmap stays current and reflects the latest information. Encourage your team to view change as an opportunity for growth and improvement rather than a setback. This mindset can foster an agile and responsive culture that can better navigate the uncertainties and challenges of product development.

Involve stakeholders in the planning process: Engage stakeholders from various departments and roles in the roadmap planning process. This collaborative approach helps ensure everyone's perspectives and priorities are considered, promoting shared ownership and commitment to the product's success. Encourage open dialogue and feedback during planning sessions to create a roadmap that balances competing priorities while remaining flexible and adaptable.

Communicate changes effectively: When changes to the roadmap occur, communicate them clearly and promptly to all stakeholders. Be transparent about the reasons for the changes, the impact on the product's goals and objectives, and any adjustments to priorities or timelines. This open communication helps manage expectations and maintains stakeholder trust. Regularly update your stakeholders on the product's progress, highlighting successes and addressing challenges to keep everyone informed and engaged.

Leverage tools and techniques: Utilize product management tools and techniques, such as Agile methodologies or the MoSCoW prioritization method, to help balance flexibility and structure in your roadmap. These tools can help your team focus on priorities while remaining agile and responsive to change. Agile methodologies, for instance, emphasize iterative development, allowing your team to adapt and respond to changing requirements quickly. At the same time, the MoSCoW method helps prioritize features based on their importance to the product's success.

By balancing flexibility and structure in your product roadmap, you can create a dynamic, adaptable plan that enables your team to navigate the uncertainties and challenges of product development while staying aligned with the product's strategic goals and objectives. This balance is critical to ensuring your product's long-term success and stakeholders' satisfaction.

Measure and Monitor Progress

Effectively measuring and monitoring progress is critical to a successful product roadmap. Tracking progress helps you stay on course, adjust plans as needed, and maintain stakeholder transparency. In this section, we'll delve deeper into strategies for effectively measuring and monitoring your product roadmap's progress:

Establish KPIs: Define Key Performance Indicators (KPIs) that align with your product's strategic goals and objectives. Ensure these KPIs are measurable, relevant, and actionable, providing insights into your product's performance and progress. Examples of KPIs include user acquisition, retention, revenue growth, and customer satisfaction.

- Track KPIs using relevant tools, such as analytics platforms or custom-built dashboards.
- Set targets for each KPI to ensure they remain ambitious yet achievable.
- Communicate KPIs and targets with your team and stakeholders to ensure alignment and shared understanding.

Set milestones: Identify critical milestones along your product roadmap, marking significant achievements or points in your product's development. Milestones can include completing important features, hitting key metrics, or reaching specific deadlines. These milestones provide a structure for monitoring progress and help keep your team focused on achieving your product's goals.

- Break down milestones into smaller tasks or objectives to make them more manageable.
- Assign responsibilities to team members to ensure accountability.
- Celebrate milestones when they are achieved to maintain team morale and motivation.

Regularly review progress: Schedule regular progress reviews with your team and stakeholders to discuss your product's progress against the roadmap. These reviews can be weekly, bi-weekly, or monthly, depending on your organization's needs. Use these meetings to:

- Discuss any deviations from the plan and identify root causes.
- Address any challenges or obstacles that may be hindering progress.
- Celebrate successes and acknowledge individual and team contributions.

Use data-driven insights: Leverage data and analytics to monitor your product's progress, identify trends, and inform decision-making.

- Track your KPIs and other relevant metrics, comparing them against your goals and targets.
- Use visualization tools to make data more accessible and understandable.
- Regularly assess your data to identify areas for improvement, optimization, or innovation.

Maintain transparency: Share progress updates with your team and stakeholders regularly, maintaining transparency and fostering a sense of shared ownership in your product's development.

- Use communication channels like email, instant messaging, or internal social networks to share updates.
- Create a centralized hub or dashboard where team members and stakeholders can view progress at a glance.
- Encourage open discussion and feedback to ensure everyone remains engaged and informed.

Adjust and adapt: As you measure and monitor progress, be prepared to adjust and adapt your product roadmap as needed.

- Stay responsive to changes in the market, new opportunities, or resource constraints that may impact your plans.
- Conduct regular roadmap reviews to assess the need for adjustments or reprioritization.
- Foster a culture of flexibility and adaptability within your team, empowering them to navigate challenges and capitalize on new opportunities.

By effectively measuring and monitoring your product roadmap's progress, you can ensure that your product stays on track, make informed decisions, and maintain alignment and transparency with your team and stakeholders. This proactive approach helps you navigate the complex world of product management, ultimately driving your product's growth and success.

Incorporate Stakeholder Feedback and Expectations

Incorporating stakeholder feedback and expectations into your product roadmap is essential to ensure alignment, secure buy-in, and foster a collaborative environment that supports your product's success. Engaging with stakeholders and soliciting their input helps you create a roadmap that reflects your organization's diverse perspectives, priorities, and expertise. Below, we'll explore strategies for effectively incorporating stakeholder feedback and expectations into your product roadmap:

Identify key stakeholders: Start by identifying the key stakeholders who should be involved in the roadmap planning process. These stakeholders may include product management, engineering, design, marketing, sales, customer support, and executive leadership representatives. Be mindful of including a diverse range of perspectives to ensure a balanced and comprehensive view of your product's priorities and objectives.

Schedule regular stakeholder meetings: Establish a cadence of regular meetings with stakeholders to discuss the product roadmap, gather feedback, and address any concerns. These meetings can be held weekly, monthly, or quarterly, depending on your organization's needs and the level of involvement required from each stakeholder. Regular meetings help maintain alignment, facilitate communication, and foster a sense of shared ownership in the product's development.

Facilitate open dialogue: Encourage open and honest discussion during stakeholder meetings, allowing everyone to voice their opinions, concerns, and ideas. This transparent communication helps identify potential issues early on, enables better decision-making, and builds stakeholder trust. Create an inclusive and respectful environment where all stakeholders feel comfortable sharing feedback.

Prioritize stakeholder input: When gathering stakeholder feedback, prioritize input based on its relevance to the product's strategic goals and objectives. This prioritization helps address and incorporate the most critical feedback into the roadmap. Be transparent about the prioritization process, explaining the rationale behind your decisions to maintain stakeholder trust and buy-in.

Document and track feedback: Keep a record of the feedback received from stakeholders and any associated decisions or action items. Share the documented feedback and any updates with stakeholders to ensure transparency and maintain engagement. This documentation helps maintain accountability, track progress and provides a reference point for future roadmap planning.

Manage expectations: As you incorporate stakeholder feedback into your product roadmap, be mindful of managing expectations regarding timelines, resources, and priorities. Communicate any trade-offs or constraints that may impact the implementation of certain features or initiatives. Setting realistic expectations and maintaining open communication can foster trust and confidence in the product roadmap among stakeholders.

By incorporating stakeholder feedback and expectations into your product roadmap, you can create a comprehensive, well-informed plan that reflects your organization's collective knowledge and priorities. This collaborative approach helps ensure the long-term success of your product and fosters a supportive environment where everyone is invested in the product's growth and development.

Be transparent about roadmap uncertainties and risks

No product roadmap is perfect, and there will always be uncertainties and risks that can impact your product's development. Being transparent about these factors with your team and stakeholders is essential to managing expectations and facilitating open discussions around potential challenges and mitigation strategies.

Identify potential risks: During the roadmap creation process, work with your team to identify and document potential risks associated with each initiative, such as resource constraints, dependencies, technical challenges, or market uncertainties.

Assign probabilities and impact levels: For each identified risk, assign a probability (low, medium, high) and an impact level (minor, moderate, major) to help prioritize which risks require the most attention and planning.

Develop contingency plans: For high-priority risks, create contingency plans that outline your team's steps to mitigate or address the potential challenges. This could include identifying alternative solutions, reallocating resources, or adjusting timelines.

Communicate risks and uncertainties openly: When presenting your roadmap to stakeholders, discuss the identified risks and uncertainties, and share your contingency plans. This transparency will help create a shared understanding of the potential challenges ahead and foster a collaborative approach to addressing them.

By following these guidelines, you'll be able to create a product roadmap that is a valuable tool for planning and prioritization and a means of fostering collaboration, alignment, and shared understanding among your team and stakeholders. Remember, a successful product roadmap is a living document that evolves with your product and organization, and maintaining its relevance and effectiveness requires ongoing attention, communication, and adjustment.

Key Takeaways

- Ensure the roadmap aligns with product vision and strategy for consistency and focus.
- Collect feedback and use prioritization frameworks to identify and rank features and goals.
- Collaborate with cross-functional teams to refine the roadmap and iterate based on feedback.
- Select the appropriate roadmap type based on product stage, stakeholder needs, and adaptability.
- Balance flexibility and structure to accommodate changes and maintain focus.
- Define KPIs, set milestones, and regularly review progress using data and analytics.
- Incorporate stakeholder feedback, schedule meetings, and manage expectations.
- Be transparent about roadmap uncertainties and risks, and develop contingency plans.
- Regularly adjust and adapt the roadmap based on market changes and resource constraints.

Chapter 18

Ensuring a Smooth Ride: Updating and Maintaining Your Roadmap

"The measure of intelligence is the ability to change." - Albert Einstein

In the ever-evolving world of product management, it's essential to recognize that your product roadmap isn't set in stone. As you sail through uncharted waters, you'll inevitably encounter new challenges, opportunities, and changes in the market. To ensure your product stays on course, keeping your roadmap updated and maintaining its relevance in the face of shifting conditions is crucial.

This section will explore the importance of regularly updating and maintaining your product roadmap. We'll discuss best practices for reviewing and adjusting your roadmap, incorporating new information and feedback, and communicating changes to your team and stakeholders. By mastering these skills, you'll be better prepared to navigate the unpredictable seas of product management and keep your product on track toward success.

Regularly Review and Update Your Roadmap

To ensure that your product roadmap remains aligned with your product vision and strategy, it's essential to establish a routine of regular reviews and updates. By revisiting your roadmap periodically, you can assess its ongoing relevance, make necessary adjustments, and ensure that it continues to effectively guide your product's development. Here are some tips for conducting regular roadmap reviews:

Establish a review cadence: Determine a suitable frequency for reviewing your roadmap, whether monthly, quarterly, or at another interval that works for your organization. The cadence should be based on the pace of your industry, the complexity of your product, and the level of uncertainty in your market.

Involve key stakeholders: Invite representatives from different teams, such as engineering, design, marketing, and sales, to participate in roadmap reviews. This ensures that diverse perspectives are considered and helps maintain alignment across the organization.

Evaluate progress and alignment: During the review, assess the progress made towards achieving the goals and objectives outlined in the roadmap. Also, evaluate if the roadmap remains aligned with your product vision and strategy.

Adjust and reprioritize: Based on your evaluation, identify necessary adjustments to the roadmap, such as reprioritizing features, reassigning resources, or reevaluating timelines. Be prepared to make tough decisions to maintain focus on your product vision and strategic goals.

Document changes and decisions: Keep a record of the changes made to the roadmap during the review process, along with the rationale for each decision. This will help maintain transparency and facilitate future roadmap reviews.

Incorporate New Information and Changing Priorities

As you navigate the dynamic landscape of product management, new information, and changing priorities will inevitably arise. To ensure your roadmap remains effective, it's crucial to incorporate this new data and adjust your plans accordingly. Here are some tips for handling changes in information and priorities:

Monitor industry trends and market changes: Stay informed about emerging trends, competitor activities, and changes in the market landscape that may impact your product's development. Be prepared to adjust your roadmap based on this information.

Gather feedback from customers and users: Continuously collect feedback from your customers and users to better understand their needs and pain points. Use this feedback to refine your product's features, prioritize enhancements, and identify new opportunities.

Track performance metrics: Regularly monitor key performance indicators (KPIs) and other metrics to assess the success of your product and its features. Use this data to inform your roadmap decisions and prioritize initiatives based on their impact and effectiveness.

Embrace change as an opportunity: Instead of resisting change, embrace it as an opportunity for growth and improvement. Recognize that updating your roadmap in response to new information or changing priorities can lead to a stronger and more successful product.

Communicate changes proactively: When you make adjustments to your roadmap, be proactive in communicating these changes to your team and stakeholders. Explain the rationale behind the changes and the expected impact on the product's development to maintain buy-in and support.

Communicate Roadmap Updates to Stakeholders

Keeping stakeholders informed about changes to your product roadmap is crucial for maintaining alignment, support, and trust within your organization. Effective communication ensures that everyone understands the rationale behind the updates and is aware of potential impacts on their work. Here are some tips for communicating roadmap updates to stakeholders:

Be transparent: Share the reasoning behind any changes to the roadmap, including the new information or shifting priorities that prompted the updates. This transparency helps to build trust and demonstrates that your decisions are based on sound reasoning.

Choose the proper communication channels: Use a variety of communication channels to share roadmap updates with stakeholders, such as meetings, presentations, email updates, or internal collaboration tools. Select the channels that best suit the needs and preferences of your audience.

Tailor your message: Adjust your communication style and level of detail to suit the needs of different stakeholder groups. For example, technical teams may require more detailed information about feature changes, while executive stakeholders may prefer a high-level summary of the impact on strategic goals.

Invite feedback: Encourage stakeholders to provide input and ask questions about the roadmap updates. This open dialogue can help address any concerns, identify potential issues, and uncover new opportunities for improvement.

Establish a cadence for communication: Develop a routine for sharing roadmap updates with stakeholders, whether on a monthly, quarterly, or ad-hoc basis. This regular communication helps to keep everyone informed and aligned with the product's direction.

Embrace Continuous Improvement

A product roadmap is not a static document; it's a living, evolving plan that should be continuously improved based on new information, feedback, and lessons learned. Embracing a mindset of continuous improvement can help your product roadmap remain effective and relevant in the ever-changing world of product management. Here are some tips for incorporating continuous improvement into your roadmap management process:

Learn from successes and failures: Reflect on your product's past successes and failures to identify areas for improvement.

Use these insights to inform your roadmap decisions and guide future product iterations.

Encourage a culture of experimentation: Foster a culture that values experimentation, learning, and adaptation. Encourage your team to test new ideas, gather feedback, and iterate on features to drive continuous improvement and innovation.

Measure and analyze: Use data and analytics to measure the effectiveness of your roadmap initiatives and inform your decision-making. Regularly review key performance indicators (KPIs) and other metrics to assess progress and identify areas for improvement.

Iterate and adapt: Be open to changing your roadmap based on feedback, new information, and changing priorities. Recognize that your product roadmap is a dynamic plan that should evolve alongside your product and its market.

Share best practices: Foster knowledge sharing and collaboration within your organization by sharing best practices, lessons learned, and success stories related to roadmap management. This can drive continuous improvement and elevate the overall effectiveness of your product management process.

Key Takeaways

- Product roadmaps should be flexible and updated regularly based on industry changes, product feedback, and progress towards goals.
- Establish a review frequency that suits your industry and product, involving key stakeholders in these reviews for diverse input and alignment.
- Document changes to the roadmap, explaining the reasoning behind each adjustment.
- Monitor industry trends, collect user feedback, and track KPIs to inform your roadmap decisions.
- Changes to the roadmap should be viewed as growth opportunities, and these changes should be proactively communicated to the team and stakeholders.
- Keep stakeholders informed about roadmap changes through transparent communication, tailoring the message to different groups, and encouraging their feedback.
- Foster a culture of continuous improvement, learning from past outcomes, promoting experimentation, and using data-driven insights to guide roadmap revisions.

Chapter 19

Illustrating Your Journey: Visualizing and Showcasing Your Product Roadmap

"The art of communication is the language of leadership." - James Humes

A well-designed product roadmap visualization can be a powerful tool for communicating your product's direction, progress, and priorities to stakeholders. It helps to make complex information more accessible and engaging, allowing your audience to grasp the key elements of your plan quickly. In this section, we'll cover the following aspects of visualizing and presenting your product roadmap:

Selecting the Right Visualization Format: Choose a visualization format that aligns with your roadmap type, audience, and communication goals. Standard designs include Gantt charts, timeline views, Kanban boards, and swimlane diagrams. We'll discuss the advantages and use cases for each format.

Designing for Clarity and Impact: Learn best practices for creating a visually appealing and easy-to-understand roadmap. We'll cover using color, typography, and layout to convey information effectively and make your roadmap stand out.

Tailoring Your Roadmap for Different Audiences: Discover how to customize your roadmap presentation for different stakeholder groups, emphasizing the most relevant information to their needs and interests.

Telling a Compelling Story with Your Roadmap: Explore storytelling techniques that can help you present your roadmap in a way that engages your audience, highlights your product's unique value, and builds support for your vision and strategy.

Leveraging Roadmap Tools and Software: Review some of the popular product roadmap tools available and learn how they can help you create, maintain, and share your roadmap more efficiently.

By the end of this section, you'll have the knowledge and insights needed to create a visually compelling product roadmap that effectively communicates your product's direction and priorities to stakeholders.

Selecting the Right Visualization Format

Choosing the proper visualization format for your product roadmap is essential to communicate your product's direction and priorities to stakeholders clearly. Different visualization formats can better suit specific types of roadmaps, audiences, and communication goals. This section will explore four common visualization formats and discuss their advantages and use cases to help you select the correct format for your roadmap.

Gantt Charts

A Gantt chart is a horizontal bar chart representing a project schedule over time. It shows activities, tasks, or features alongside a timeline, illustrating their start and end dates, duration, and dependencies.

Advantages:

- Provides a clear overview of your project timeline, milestones, and deadlines.
- Highlights dependencies between tasks and potential bottlenecks.
- Easy to understand and widely recognized.

Use Cases:

- Time-based or feature-based roadmaps.
- Ideal for projects with a strong focus on schedules and deadlines.
- Suitable for both internal and external stakeholders.

Example:

A software development team uses a Gantt chart to visualize the release schedule for a new application, including features, milestones, and dependencies between tasks.

Timeline Views

A timeline view is a linear representation of events or activities arranged chronologically. It shows the sequence and relative timing of roadmap items, such as goals, features, or themes, along a horizontal or vertical axis.

Advantages:

- Gives a high-level overview of the product's development journey.
- Emphasizes the sequence and timing of roadmap items.
- Flexible and adaptable to different roadmap types.

Use Cases:

- Goal-oriented, feature-based, or theme-based roadmaps.
- Ideal for presenting a high-level view of your product's development plan.
- Suitable for a wide range of audiences, from executives to customers.

Example:

A product manager creates a timeline view of a goal-oriented roadmap to show the progression of strategic objectives over the next two years.

Kanban Boards

A Kanban board is a visual workflow management tool that represents tasks or activities as cards on columns or lanes, corresponding to different stages of the process. It helps track progress and identify bottlenecks in real time.

Advantages:

- Offers a visual representation of your product's workflow and progress.
- Encourages collaboration and transparency within your team.
- Facilitates agile and lean project management prac-

tices.

Use Cases:

- Suitable for internal team collaboration and communication.
- Works best with agile development methodologies and shorter-term planning.
- It can be adapted to represent various feature- or theme-based roadmap types.

Example:

A product team uses a Kanban board to manage their feature development process, with columns representing stages like "Backlog," "In Progress," and "Completed."

Swimlane Diagrams

A swimlane diagram is a flowchart that organizes tasks or activities into parallel lanes, representing different categories or groups, such as teams, departments, or priorities.

Advantages:

- Highlights the responsibilities and collaboration between different teams or groups.
- Shows the relationship and dependencies between

roadmap items across categories.

- Provides a more structured view of your product's development plan.

Use Cases:

- Ideal for cross-functional projects and initiatives.
- Works well with goal-oriented, feature-based, or theme-based roadmaps.
- Suitable for internal communication and collaboration.

Example:

A product manager creates a swimlane diagram to visualize the development plan for a new feature, with lanes representing different teams, such as design, development, and marketing.

By understanding these visualization formats' unique characteristics, advantages, and use cases, you can select the one that best aligns with your roadmap type, audience, and communication goals. The right design will make your roadmap more accessible, engaging, and informative, ensuring stakeholders can quickly grasp the essential elements.

Designing for Clarity and Impact

Creating a visually appealing and impactful product roadmap is crucial for effectively communicating your product's direction and priorities to stakeholders. A well-designed roadmap should be clear, concise, and visually engaging, making it easy for your audience to understand and digest the information. This section will discuss several design principles and tips to help you create a roadmap with clarity and impact.

Choose a clean and straightforward layout: Opt for a design that is easy to understand and navigate. Avoid clutter and unnecessary elements that may distract from the main message. A clean and straightforward layout makes it easier for your audience to quickly focus on the essential information and grasp the key points.

Use consistent color coding and visual cues: Assign specific colors and visual cues to represent different categories, priorities, or status levels. For example, you can use different colors for high, medium, and low-priority features or other teams and departments. Compatible color coding and visual cues help your audience quickly recognize and interpret the information on your roadmap.

Highlight critical milestones and achievements: Emphasize essential milestones, such as product launches, major releases, or strategic goals, by making them stand out from other roadmap items. This can be achieved using bold colors, icons, or larger text sizes. Highlighting key milestones helps your audience understand the significant events and achievements on your product's journey.

Utilize whitespace effectively: Whitespace, or the empty space between elements on your roadmap, is essential for readability and clarity. Proper use of whitespace helps separate different sections, groups, or categories, making it easier for your audience to follow and understand the structure of your roadmap.

Choose an appropriate font size and style: Select a font size and style that is easy to read and understand. Avoid overly decorative or stylized fonts that may be difficult to read or interpret. Choose a font size large enough to be legible, especially when presenting your roadmap on a large screen or printed format.

Use clear labels and annotations: Ensure your roadmap items, such as goals, features, or milestones, have clear and concise labels. Clear labels and annotations help your audience quickly grasp the meaning and relevance of each roadmap item. Additionally, provide annotations or tooltips to offer additional context or explanations when necessary.

Optimize for different devices and formats: Consider how your roadmap will be viewed on other devices, such as laptops, tablets, or smartphones, and ensure it is responsive and adaptable to various screen sizes and resolutions. Additionally, ensure your roadmap can be easily printed or exported to different formats, such as PDF or PowerPoint, for presentation and sharing purposes.

By following these design principles and tips, you can create a product roadmap that is visually appealing, clear, and impactful. A well-designed roadmap will help you communicate your product's direction and priorities effectively and engage and inspire your stakeholders, leading to more significant support and collaboration for your product's success.

Tailoring Your Roadmap for Different Audiences

Understanding that different audiences have varying needs and expectations regarding your product roadmap is essential. By tailoring your roadmap to suit each audience's specific interests and concerns, you can ensure your message is clear, relevant, and engaging. This section will discuss strategies for adapting your product roadmap for different audiences, including internal teams, executives, and external stakeholders.

Internal teams: When presenting your roadmap to internal groups, such as developers, designers, or marketing professionals, focus on the details that are most relevant to their specific roles and responsibilities. This may include feature specifications, technical requirements, design mockups, or marketing initiatives. Make sure to highlight the dependencies between teams and the expected outcomes of each team's efforts. Encourage open communication and collaboration, and provide opportunities for feedback and input.

Executives and senior management: When presenting your roadmap to executives and senior management, prioritize high-level strategic goals, key performance indicators (KPIs), and the overall progress toward achieving your product vision. Emphasize the alignment of your roadmap with the company's strategic objectives and demonstrate how your product contributes to the organization's success. Keep your presentation concise and focused on critical information, as executives often have limited time and attention.

External stakeholders: When presenting your roadmap to external stakeholders, such as investors, partners, or customers, focus on the benefits and value that your product brings to them. Highlight upcoming features, improvements, or initiatives that address their specific pain points, needs, or interests. Be mindful of confidentiality concerns and avoid sharing sensitive or proprietary information that could jeopardize your competitive advantage.

Adapting your visualization: Depending on your audience, you may need to adjust the level of detail, format, or visual style of your roadmap. For example, a detailed, feature-based roadmap may be appropriate for internal teams, while a high-level, goal-oriented roadmap might be more suitable for executives. Choose the best visualization format to convey the most relevant and valuable information to each audience.

Prepare for questions and feedback: Anticipate the questions and concerns that different audiences may have about your product roadmap and be prepared to address them. This may involve providing additional context, data, or examples to support your decisions and priorities. Encourage open dialogue and feedback, as this can help you identify potential risks, opportunities, or areas for improvement in your roadmap.

By tailoring your product roadmap for different audiences, you can ensure your message is clear, relevant, and engaging to each stakeholder group. This approach fosters better communication, collaboration, and support for your product's development and success, ultimately contributing to achieving your product vision and strategic goals.

Telling a Compelling Story with Your Roadmap

A well-crafted product roadmap is more than just a collection of features, dates, and priorities; it's also a powerful storytelling tool that can help you convey your product's vision, progress, and value to various stakeholders. Turning your roadmap into a compelling narrative can create a deeper connection with your audience, evoke emotions, and inspire action. This section will explore strategies for telling a compelling story with your product roadmap.

Start with your product vision: Begin your roadmap presentation by revisiting your product vision and explaining how the roadmap serves as a strategic plan to achieve that vision. This helps set the stage and remind your audience of your product's goals and direction.

Highlight critical milestones and achievements: As you walk through your roadmap, emphasize the significant milestones, accomplishments, and learnings along the way. By showcasing your product's progress and successes, you can instill confidence in your audience and demonstrate your team's ability to execute your vision.

Connect the dots: To create a cohesive narrative, show the logical flow and interdependencies between various roadmap items. Explain how each feature, initiative, or goal contributes to the product vision and strategic objectives. This helps your audience understand the rationale behind your prioritization and decision-making.

Use relatable analogies and examples: To make your roadmap story more engaging and memorable, use analogies or real-life examples to illustrate key points or concepts. This can help your audience better grasp complex ideas and create a more vivid mental picture of your product's future state.

Share customer stories and feedback: Incorporate customer stories, testimonials, or feedback into your roadmap narrative to demonstrate the real-world impact of your product. This adds credibility to your story, helps humanize your product, and creates an emotional connection with your audience.

Be transparent about challenges and risks: Don't avoid discussing the challenges, risks, or uncertainties you've encountered or anticipate in your product's journey. Being open and honest about these issues can foster trust and credibility with your audience and demonstrate your commitment to continuous improvement and learning.

End on a high note: Conclude your roadmap story by reiterating your product vision and the progress made toward achieving it. Highlight the opportunities and exciting developments ahead, and express confidence in your team's ability to deliver on your roadmap commitments.

By transforming your product roadmap into a compelling story, you can create a more engaging and impactful presentation that resonates with your audience. This storytelling approach helps foster better understanding, buy-in, and support for your product's development and success, ultimately contributing to realizing your product vision and strategic goals.

Leveraging Roadmap Tools and Software

In today's fast-paced, technology-driven world, plenty of tools and software are designed to help you create, manage, and share your product roadmap. These solutions can streamline your roadmap process, enhance collaboration, and enable you to present your roadmap in a visually appealing and easily digestible format. In this section, we'll discuss some key considerations when selecting and leveraging roadmap tools and software and provide examples of popular solutions along with their pros and cons.

Identify your needs and requirements: Before choosing a roadmap tool or software, take the time to understand your specific needs and requirements. Consider factors such as your roadmap type, the complexity of your product, the size of your team, and the preferred level of customization and flexibility.

Evaluate features and functionality: As you explore various roadmap tools and software, evaluate their features and functionality to ensure they align with your needs. Some essential features include drag-and-drop capabilities, timeline views, prioritization tools, progress tracking, and integration with other project management or communication tools.

Collaborative capabilities: Choose a roadmap tool that facilitates collaboration and communication among your team members and stakeholders. Features such as real-time updates, commenting, and version control can ensure everyone stays on the same page and can contribute effectively to the roadmap process.

Customizability and scalability: As your product and organization grow, your roadmap needs may change. Select a tool that offers a high degree of customization and scalability, allowing you to adapt your roadmap as your requirements evolve.

Security and data protection: Given the sensitive nature of roadmap information, choosing a tool that provides robust security measures and data protection capabilities is crucial. Ensure the tool adheres to industry-standard security practices and offers features such as data encryption and access controls.

Assess ease of use and learning curve: When selecting a roadmap tool or software, consider its ease of use and associated learning curve. Opt for an intuitive and user-friendly solution, enabling your team to get up to speed and maximize productivity quickly.

Cost and value for money: Lastly, consider the cost of the roadmap tool or software and whether it offers value for money. Compare different solutions and pricing plans to find one that meets your budget and provides the needed features and capabilities.

By carefully selecting and leveraging the right roadmap tools and software, you can streamline your roadmap process, improve collaboration, and present your product roadmap in a visually compelling and easily digestible format. This, in turn, can help you better communicate your product vision and strategy to your team and stakeholders, ultimately driving your product's success.

Consider your specific needs and requirements when choosing a roadmap tool or software. Evaluate each solution based on features, functionality, ease of use, collaboration capabilities, customizability, scalability, security, and cost. Doing so will enable you to manage your product roadmap effectively and ensure it aligns with your product vision and strategy.

Once you've selected the right tool for your needs, take the time to learn its features and capabilities and train your team to use it effectively. Encourage open communication and collaboration through the tool, ensuring everyone is aligned and working together toward your product's success.

Remember that no single tool or software will be perfect for every organization, so it's essential to remain adaptable and open to change as your product and organization evolve. By continuously evaluating and refining your roadmap process and toolset, you'll be well-prepared to navigate the ever-changing landscape of product management and drive your product's growth and success.

Key Takeaways

- Pick the visualization format (Gantt Charts, Timeline Views, Kanban Boards, Swimlane Diagrams) for your roadmap based on your roadmap type, audience, and goals.
- Design with a clean layout, consistent color coding, highlighted milestones, and optimized for different devices.
- For internal teams, focus on details and communication. For executives, emphasize strategic goals and progress. For external stakeholders, highlight product value with confidentiality.
- Present product vision, milestones, logical flow, and challenges. Use analogies and customer stories. Conclude with progress and vision.
- Identify needs before choosing a tool. Consider features, ease of use, collaboration, customization, security, and cost. Learn the tool, encourage collaboration, and remain adaptable. Continuously refine your process and tools.

Chapter 20

Navigating Roadblocks: Overcoming Roadmap Challenges

"Success is stumbling from failure to failure with no loss of enthusiasm." - Sir Winston Churchill

As with any aspect of product management, creating, maintaining, and presenting a product roadmap comes with its fair share of challenges. In this section, we'll explore some of the most common issues that product managers face when working with roadmaps and provide actionable strategies for overcoming these obstacles. By addressing these challenges head-on, you'll be better equipped to ensure your roadmap remains valuable for guiding your product's development and aligning your team and stakeholders.

The topics we'll cover in this section include:

- Managing stakeholder expectations
- Balancing competing priorities

- Addressing uncertainty and risk
- Aligning cross-functional teams
- Ensuring roadmap flexibility
- Keeping the roadmap up-to-date

As we explore these challenges, keep in mind that overcoming them requires a combination of effective communication, collaboration, and adaptability. By embracing these qualities and learning from the insights and strategies shared in this section, you'll be well on your way to creating a roadmap that drives your product's success and helps you navigate the complex world of product management.

Managing Stakeholder Expectations

Managing stakeholder expectations is crucial to creating and maintaining a product roadmap. Stakeholders, including executives, team members, and customers, have varying perspectives and priorities regarding the product's development. Balancing these expectations and ensuring everyone is on the same page can be challenging. Here are some strategies to help you effectively manage stakeholder expectations, with additional details on each point:

Early and frequent communication: Engage stakeholders from the beginning of the roadmap process, and keep them informed throughout its development. Schedule regular meetings or touchpoints to discuss the roadmap and provide status updates through email, internal communication tools, or dedicated project management platforms. By maintaining open lines of communication, you can better manage expectations and prevent misunderstandings.

Set realistic expectations: Be honest about what can be achieved within the given time frame and resources. Avoid overpromising and underdelivering, as this can lead to disappointment and a loss of trust. Instead, set realistic expectations and be transparent about potential risks and uncertainties, explaining how they might impact the roadmap. Present a clear, data-driven assessment of the product's current state, and provide realistic estimates for feature delivery and milestone completion.

Prioritize and align: Prioritize features and initiatives based on their importance and alignment with the product vision and strategy. Develop a prioritization framework, such as a scoring system or a prioritization matrix, to help you make objective decisions. Be prepared to explain your reasoning and engage in constructive discussions with stakeholders who may have different priorities. This will help ensure that everyone understands the rationale behind your decisions and supports the direction of the product.

Establish clear roles and responsibilities: Ensure everyone involved in the product development process understands their roles and responsibilities. Create a clear organizational chart, document key responsibilities, and regularly review these with stakeholders to ensure clarity and clarity. By clarifying expectations around roles and responsibilities, you can ensure stakeholders feel engaged and accountable for their part in the process.

Demonstrate progress and success: Regularly showcase the progress made on the roadmap and any key milestones or achievements. Use visual aids like burndown charts, progress bars, or status dashboards to illustrate progress and the remaining work. Celebrate successes, even small ones, to help build stakeholder confidence in your product management capabilities and the overall direction of the product.

Be receptive to feedback: Encourage stakeholders to provide input on the roadmap and be open to adjusting your plans based on their input. Establish regular feedback channels, such as surveys, focus groups, or one-on-one meetings, to gather insights and suggestions. By fostering an environment of collaboration and continuous improvement, you can better manage stakeholder expectations and ensure the success of your product.

By proactively addressing stakeholder expectations and maintaining open lines of communication, you can create a more positive and collaborative environment for product development. This will ultimately lead to a more effective product roadmap and a greater likelihood of achieving your product's goals.

Balancing Competing Priorities

Striking the right balance is crucial to ensure the product's success and maintain stakeholder satisfaction. In creating and maintaining a product roadmap, product managers often face balancing competing priorities from various stakeholders. These priorities may include feature requests, bug fixes, technical debt, and strategic initiatives. Here are some strategies to help you effectively balance competing priorities:

Establish a clear prioritization framework: Develop a prioritization framework to objectively assess the importance and urgency of each request or initiative. This framework could be based on factors such as business impact, customer value, technical complexity, and alignment with the product vision and strategy. A clear framework can help you make informed decisions and provide a solid rationale when discussing priorities with stakeholders.

Involve stakeholders in the prioritization process: Engage stakeholders in prioritizing features, bug fixes, and other initiatives. This can help ensure their perspectives are considered and foster a sense of ownership and buy-in. Involving stakeholders in decision-making can lead to more informed decisions and a greater understanding of the trade-offs involved.

Communicate the rationale behind decisions: Communicate the reasoning behind your prioritization decisions to stakeholders. Explain how each decision aligns with the product vision, strategy, and goals and the expected impact on the product's success. By being transparent about your decision-making process, you can help stakeholders understand the trade-offs and the reasoning behind your choices.

Set boundaries and be prepared to say no: As a product manager, it's essential to set boundaries and say no to requests that do not align with the product's goals or vision. Be firm but also empathetic and understanding when communicating your decisions. Explain the reasons behind your decision, and offer alternative solutions or compromises when appropriate.

Regularly review and adjust priorities: Continuously monitor and reassess the product's priorities based on new information, changing market conditions, and stakeholder feedback. Regularly review your priorities and roadmap with stakeholders to ensure alignment and maintain buy-in. Be prepared to adjust your roadmap and reprioritize initiatives as needed to ensure your product remains on track to achieve its goals.

Maintaining a balance between short-term and long-term goals: It's crucial to address immediate needs (such as bug fixes or customer requests) and focus on strategic, long-term initiatives. Ensure that your product roadmap reflects this balance and that adequate resources are allocated to both short-term and long-term priorities. Regularly review your progress and adjust your focus as needed to maintain this balance.

By effectively balancing competing priorities, you can ensure that your product roadmap remains aligned with your product vision and strategy while addressing the needs and expectations of stakeholders. This balance will help you drive your product's success and maintain strong relationships with stakeholders throughout the product development process.

Addressing Uncertainty and Risk

Product roadmaps, by nature, involve a degree of uncertainty and risk as they chart the course for a product's future development. As a product manager, it's essential to acknowledge and address these uncertainties and risks to ensure your product's success and maintain stakeholder confidence. Here are some strategies for effectively managing uncertainty and risk in your product roadmap:

Embrace a flexible and iterative approach: Recognize that your product roadmap is a living document that will evolve over time as new information becomes available and priorities change. Adopt a flexible and iterative approach to roadmap planning, being prepared to adapt and adjust as needed. This mindset can help you navigate uncertainty and ensure your product remains on track to achieve its goals.

Identify and assess potential risks: Proactively identify risks that could impact your product's development and success, such as technical challenges, market changes, or resource constraints. Assess each risk's likelihood and potential impact, and develop contingency plans or mitigation strategies to address them.

Communicate uncertainties and risks to stakeholders: Be transparent about the uncertainties and risks inherent in your product roadmap, and communicate them clearly to stakeholders. This openness can help build trust and manage expectations, as stakeholders will understand the factors that could influence the product's development and timeline.

Involve stakeholders in risk management: Engage stakeholders in identifying, assessing, and addressing potential risks. This collaborative approach can lead to more informed decision-making and a greater understanding of the challenges and uncertainties involved in product development.

Monitor and track risks: Continuously monitor and track identified risks and any new ones that emerge during product development. Regularly review and update your risk assessments, and adjust your roadmap and mitigation strategies as needed to address changing circumstances.

Learn from past experiences: Reflect on previous projects and product development experiences to identify lessons learned and best practices for managing uncertainty and risk. Apply these insights to your current product roadmap to improve your risk management processes and increase the likelihood of your product's success.

By acknowledging and addressing uncertainty and risk in your product roadmap, you can build resilience, maintain stakeholder confidence, and ensure your product remains on track to achieve its goals. Embracing a proactive and collaborative approach to risk management can help you navigate the challenges and uncertainties inherent in product development, setting your product up for success.

Aligning Cross-Functional Teams

Aligning cross-functional teams is crucial to the success of your product roadmap. When aligned, teams work together effectively towards a common goal, leading to better communication, collaboration, and product success. Here are some strategies to help align cross-functional teams around your product roadmap:

Develop a shared understanding of the product vision and strategy: Ensure that all team members clearly understand the product vision and strategy and their role in achieving it. Communicate the vision and strategy effectively, using workshops, presentations, or other methods to engage and inform your teams.

Involve cross-functional teams in roadmap creation: Engage cross-functional teams in creating and refining your product roadmap. This collaborative approach ensures that each team's perspective and expertise are considered, leading to a more comprehensive and effective roadmap.

Set clear goals and expectations: Define clear, measurable goals and expectations for each team, and ensure that these goals align with the overall product roadmap. Establish a shared understanding of success and encourage a culture of accountability.

Foster open communication and collaboration: Encourage open communication and collaboration between cross-functional teams, breaking down silos and promoting a culture of shared ownership. Use tools and processes that support collaboration, such as regular meetings, shared documents, or project management software.

Establish a straightforward decision-making process: Define a clear and transparent process involving relevant stakeholders from across the organization. This process should balance the need for input and collaboration with efficiency and timely decision-making.

Monitor and measure progress: Regularly review and measure progress against your product roadmap, and share updates with cross-functional teams. Celebrate successes and milestones while also addressing challenges and adjusting plans as needed.

Provide ongoing support and resources: Ensure cross-functional teams have the resources and support to execute the product roadmap effectively. This may include training, tools, or additional resources to help teams overcome challenges and achieve their goals.

By implementing these strategies, you can align cross-functional teams around your product roadmap, fostering a culture of collaboration and shared ownership that drives product success.

Ensuring Roadmap Flexibility

Flexibility is a crucial aspect of effective product roadmaps. As the product, market, and organizational landscape evolve, your roadmap must adapt to accommodate new information and changing priorities. Below are several strategies to help you ensure roadmap flexibility:

Adopt an Agile mindset: Embrace the principles of Agile product management, which emphasize iterative development, continuous improvement, and responsiveness to change. This approach lets you quickly adapt your roadmap to new information and shift priorities.

Use rolling-wave planning: Instead of creating a fixed roadmap for an extended period, use rolling-wave planning to break your product development into shorter time horizons. This way, you can regularly reassess and adjust your plans based on the latest information.

Emphasize goals over specific features: Focus on the overall goals and objectives you aim to achieve rather than getting too attached to particular features. By concentrating on outcomes, you'll be more open to adjusting your plans and finding alternative solutions to reach those goals.

Create a buffer for unexpected changes: Include buffer time in your roadmap to accommodate unforeseen changes, such as new priorities, urgent bug fixes, or changing market conditions. This will give you some leeway to adapt your plans without disrupting the entire roadmap.

Regularly review and update your roadmap: Set aside time on a regular basis (e.g., monthly or quarterly) to check your roadmap and make any necessary adjustments. This ensures that your roadmap stays up-to-date and aligned with your product vision and strategy.

Communicate the flexible nature of the roadmap: Make sure your stakeholders understand that the roadmap is a living document and subject to change. This helps manage expectations and reinforces the notion that the roadmap is a guide rather than a strict, unchangeable plan.

By implementing these strategies, you'll create a more adaptable and resilient product roadmap that can effectively navigate the ever-changing landscape of product management.

Keeping the Roadmap Up to Date

A product roadmap is a living document that evolves as new information, priorities, and opportunities arise. Keeping your roadmap up to date ensures that it remains a valuable tool for guiding your product's development and maintaining alignment across your organization. Here are some tips for keeping your roadmap current and relevant:

Schedule regular roadmap reviews: Establish a regular cadence for reviewing and updating your roadmap, such as monthly or quarterly. This helps you stay on top of your product, market, and organizational changes and make timely adjustments to your plans.

Stay informed about market and industry trends: Monitor market trends, competitor activities, and emerging technologies to ensure your roadmap reflects the most current and relevant information. Regularly engage with industry news, research reports, and other sources of market intelligence.

Gather feedback from stakeholders: Regularly solicit input from key stakeholders, such as customers, team members, and executives, to ensure your roadmap remains aligned with their needs and expectations. This feedback can help you identify areas for improvement and uncover new opportunities for your product.

Monitor product performance and progress: Track key performance indicators (KPIs) and other metrics to assess the success of your product and the progress of your roadmap initiatives. Use this data to inform your roadmap updates and prioritize future efforts.

Communicate roadmap updates to stakeholders: When you make significant updates, communicate those changes to your stakeholders to maintain transparency and alignment. This can be done through meetings, presentations, or other communication channels that work best for your organization.

Be prepared to pivot: Recognize that your roadmap may need to change in response to unforeseen events or new opportunities. Be willing to pivot your plans and adapt your roadmap as necessary to stay on track toward your product vision and strategic goals.

By actively managing and updating your product roadmap, you'll ensure that it remains an accurate and valuable tool for guiding your product's development and fostering alignment across your organization. Regularly revisiting and refining your roadmap will help you stay nimble and responsive to the dynamic nature of product management.

Key Takeaways

- Engage stakeholders frequently, set realistic expectations, prioritize features, define roles, celebrate progress, and adjust to feedback.
- Utilize a prioritization framework, involve stakeholders, communicate decision reasons, set boundaries, adjust priorities, and balance short and long-term goals.
- Be flexible in planning, identify risks, communicate about risks, involve stakeholders in risk management, monitor risks, and learn from past experiences.
- Foster shared understanding of the vision, involve teams in roadmap creation, set clear goals, encourage open communication, establish a decision-making process, monitor progress, and provide resources.
- Embrace Agile, use rolling-wave planning, focus on goals over features, include buffer time, regularly update roadmap, and communicate its flexibility.
- Schedule regular reviews, stay updated on trends, gather stakeholder feedback, monitor product performance, communicate updates, and be ready to pivot as necessary.

Chapter 21

Reflection: Navigating The Waters of Product Roadmapping

"By failing to prepare, you are preparing to fail." - Benjamin Franklin

Congratulations on completing the third part of your product management journey on product roadmaps! You have now gained a comprehensive understanding of product roadmaps, their importance in the product development process, and the various techniques and best practices for creating and maintaining them.

In the previous chapters on product roadmaps, we've explored:

The Purpose and Importance of Product Roadmaps: We discussed how roadmaps serve as a strategic tool to align teams, communicate plans, and track progress toward your product vision and goals.

Types of Roadmaps: We delved into the four main types of product roadmaps: goal-oriented, feature-based, time-based, and theme-based roadmaps. Each type has unique characteristics, advantages, and use cases, helping you choose the best fit for your needs.

Creating an Effective Product Roadmap: We discussed best practices for aligning your roadmap with your product vision and strategy, prioritizing features and goals, collaborating with cross-functional teams, selecting the right roadmap type, and more.

Updating and Maintaining Your Roadmap: We covered the importance of regularly reviewing and updating your roadmap, incorporating new information and changing priorities, and communicating roadmap updates to stakeholders.

Visualizing and Presenting Your Product Roadmap: We explored various visualization formats and best practices for designing a clear and impactful roadmap, tailoring it to different audiences, and leveraging tools and software.

Common Roadmap Challenges and How to Overcome Them: We examined some of the most common challenges product managers face when creating and managing roadmaps and shared strategies for overcoming them.

As you reflect on the insights and examples discussed about product roadmaps, consider how you can apply these lessons to your product management practice. Mastering the art of product road mapping will help you chart a clear and strategic course for your product's development and success.

Chapter 22

Back to Basics: Final Thoughts on Your Product Management Journey

"The only impossible journey is the one you never begin." - Tony Robbins

Congratulations on completing this introduction book to your product management journey! You now have a comprehensive understanding of the bedrocks of product management – product vision, product strategy, and product roadmap – and their critical roles in guiding your product's development and success.

Throughout your journey, we've explored the following:

Product Vision: The foundation of your product, providing inspiration, direction, focus, alignment, and differentiation. We covered crafting a powerful vision statement, aligning your team and organization, evolving your vision over time, and effectively communicating your vision to stakeholders.

Product Strategy: The bridge between your product vision and execution, helping you make informed decisions and navigate the competitive landscape. We discussed understanding your market and customers, analyzing the competition, prioritizing initiatives, setting strategic goals and objectives, and balancing short-term and long-term objectives.

Product Roadmap: A vital tool that outlines the path to achieving your product vision and strategy. We examined different types of roadmaps, creating effective roadmaps, factors contributing to an effective roadmap, updating and maintaining your roadmap, visualizing and presenting your roadmap, and addressing everyday challenges.

As you wrap up this book, take a moment to reflect on the insights and examples shared. Consider how you can apply these lessons to your product management practice. Remember, mastering the "Product Management Power Trio" is essential for navigating the dynamic world of product management and achieving success with your product.

Made in United States
North Haven, CT
04 November 2026

21769522R00129